# URBAN HYDROLOGY

M. J. HALL

*Principal Hydrologist, Sir William Halcrow and Partners, London and Swindon, UK*

ELSEVIER APPLIED SCIENCE PUBLISHERS

LONDON and NEW YORK

ELSEVIER APPLIED SCIENCE PUBLISHERS LTD
Ripple Road, Barking, Essex, England

*Sole Distributor in the USA and Canada*
ELSEVIER SCIENCE PUBLISHING CO., INC.
52 Vanderbilt Avenue, New York, NY 10017, USA

**British Library Cataloguing in Publication Data**

Hall, M. J.
Urban hydrology.
1. Hydrology
I. Title
551.48 GB661.2

ISBN 0-85334-268-7

WITH 53 ILLUSTRATIONS AND 15 TABLES



Printed in Northern Ireland at The Universities Press (Belfast) Ltd.

# Preface

The last thirty years have seen the subject of hydrology mature from being largely an engineering design aid into a recognised branch of the science of geophysics. This progression has been brought about mainly by the growth of interest from individuals drawn from disciplines other than civil engineering. As a result of their contribution, techniques such as systems analysis, stochastic processes and mathematical programming have been explored and introduced, thereby shedding new light on old problems. The catalyst for these developments has, of course, been the increasingly wide availability of computational aids, ranging from hand-held calculators to the larger mainframe computers.

Despite the achievements of the last three decades, there have also been some disturbing tendencies. Hydrology deals with natural phenomena operating over an extended range of temporal and spatial scales. Improvements in understanding tend to be made in a long series of minute steps rather than in the giant leaps symbolised by an astronaut's Moonwalk. Moreover, these steps are documented in an increasingly diverse selection of professional journals, often having a limited circulation. The result has been the opening of an ever-widening gulf between the academic and research community and the practitioners whom they purport to serve. The former accuse the latter of undue conservatism in the choice of the techniques that they apply, and the latter accuse the former of pursuing increasingly esoteric lines of investigation that lack relevance to real-world problems.

Substantial progress in resolving this apparent dichotomy could be achieved if both groups paid greater attention to presentation. The practitioner ideally requires a technique to be set out in a step-by-step format, with all relevant information between one set of covers. In contrast, the researcher is more interested in the underlying theory, the justification for approximations and the choice of key parameters. These two requirements are not readily satisfied within the same volume, and attempts to do so can result in the alienation of both groups of potential users. A manual or design guide is invariably judged by the brevity and relevance of its contents. An obvious means of achieving these qualities is to omit the basic theory, but where then can this material be presented and discussed? As the engineering problems become more complex, the need grows for a clear understanding of techniques and their limitations. When the problem involves the hydrology of a catchment whose land use is changing, such knowledge becomes a prerequisite.

With the increasing pressure on land and water resources that has accompanied the growth in the world's population, the changes in flow regime brought about by alterations in the land use of a drainage basin have inevitably attracted more attention. Nowhere are the changes more dramatic than those observed when a predominantly undisturbed rural catchment undergoes urban development. Urbanisation alters both the ground cover and the nature of the catchment drainage system and poses two distinct engineering problems: the design of the new drainage system; and the protection of the area downstream from the changes in flow regime brought about by the urban development. In addition, the urban area generates an increasing water supply demand that can rarely be satisfied from local sources, and creates further problems of waste disposal that have profound implications for the quality of local water bodies. These are the principal topics with which urban hydrology is concerned.

Although interest in urban hydrology is now worldwide, the progress that has been made in the United Kingdom has been notable. An appreciation of the need to update the current methods of stormwater drainage design for urban areas led in 1974 to the formation of the Working Party on the Hydraulic Design of Storm Sewers (HDSS). Their Report, published in 1981, not only provided a range of design methods collectively known as the Wallingford Procedure, but also introduced a fresh approach to urban drainage design that allows the performance of a sewerage system to be evaluated under conditions more onerous than the selected design standard. In addition, concern about the increased incidence of flooding downstream of drainage areas subject to urbanisation and the lack of guidance on methods for mitigating the effects of the changes in flow regime stimulated the

Construction Industry Research and Information Association (CIRIA) into initiating the preparation of a Guide to the design of flood storage ponds in 1976. Both the HDSS Report and the CIRIA Guide were based upon approaches to flood estimation similar to those promulgated in the Flood Studies Report published by the Natural Environment Research Council in 1975.

The Flood Studies Report is a comprehensive document by any standards, including as it does everything from basic data tabulations to simple step-by-step calculations, but the HDSS Report and the CIRIA Guide make fewer concessions to the non-specialist. The Wallingford Procedure may be regarded as the culmination of developments that can be traced from the middle of the nineteenth century. The flood estimation methods contained in all three documents employ an innovative approach to relate the frequency of the peak flow rate to that of the causative storm event. How do these developments compare with those proposed elsewhere? Have hydrologists in other countries approached the problems posed by urbanisation using alternative methods? Have they achieved any greater success? An examination of these and other related questions provided the initial motivation to prepare this text.

The treatment of urban hydrology would have been incomplete without considering water quality problems of urban areas. A general appreciation of the latter has been slow to develop, and the interplay between water quality and quantity, which forms the basis of successful stormwater management, appears as yet to have attracted little recognition. A review of these problems and some of the solutions that have been proposed has therefore been included in this book.

The major portion of the material upon which this text has been based was originally presented as a lecture course to postgraduate students in both Engineering Hydrology and Public Health Engineering at the Imperial College of Science and Technology in London. An extended series of mid-career training courses on various aspects of drainage design, promoted latterly by PTRC Education and Research Services Ltd, have provided the further incentive to organise its presentation. The feedback that has been obtained over the years from both students and participants has been invaluable in determining the depth of treatment for individual topics. To those individuals, including fellow members of various working parties and the writer's present colleagues in Sir William Halcrow and Partners, too many to mention by name, who have helped to shape the contents of this book, go my sincere thanks. Last, but by no means least, no acknowledgements would be complete without mention of my family and their forbearance during the preparation of the manuscript.

M. J. HALL

# Contents



## Part I: Meteorology

## Part II: Flood Hydrology



## Part III: Hydrological Problems of Urban Areas

# 1

# The Origin of Urban Hydrology

## 1.1. INTRODUCTION

Although urbanisation, which may be broadly defined as the process of expanding urban influence, has been taking place for more than 6000 years in some places, its pace has increased markedly since the beginning of the nineteenth century. Prior to that time, most urban settlements were small and functioned largely as market towns serving the surrounding countryside. In effect, the urban population was dependent upon the food surplus created by a more numerous agricultural community. Moreover, since the rural food producers would bring goods into the urban area, sell them and return to their homes between daybreak and nightfall, the radius of influence of each town was strictly limited.

The urban growth that has taken place worldwide since the early 1800s may be largely attributed to the Industrial Revolution. With the expansion of manufacturing industries, towns grew in size in order to accommodate the extra labour force. Simultaneously, mechanised agriculture became capable of generating a larger food surplus with fewer farm workers. By the second half of the nineteenth century, improvements in transportation, brought about nationally by the growth of railways and internationally by the development of the steamship, had severed the traditional dependence of urban areas on their environs for food, and the continuous increase in the ratio of urban to rural dwellers was well established.

The improvements in transportation systems which assisted the movement of food and manufactured goods also brought about a change in living habits which has resulted in the area covered by streets and buildings expanding out of all proportion to the absolute increase in the urban population. Whereas previously an urban worker would almost invariably sleep at night on the upper floor of the building whose ground floor was his place of work, cheap transportation enabled him to travel each day between his urban workplace and a surburban dormitory area. This practice of commuting not only resulted in each urban worker occupying two units of space instead of one, but also influenced the pattern of development on the outskirts of the urban area. When the majority of the working population were dependent upon public transport, such as railway and tram systems, urban areas assumed a star-shaped form, with linear development along the main road systems. The increasing dependence upon road transportation during the first half of the twentieth century has led to further dispersal. The central business area of a town or city has lost its dominance, and the urban conurbation has tended to become multi-centred. Legislation such as that which designated a Green Belt around London after the Second World War was intended specifically to limit the spread of such dispersed development. Nevertheless, the outer fringes of major cities have increasingly become broad transition zones, with adjacent metropolitan centres growing together. The most frequently quoted example of the latter is that of the north-eastern seaboard of the United States, where the urban and suburban land use stretches for some 1100 km. This area was christened Megalopolis by J. Gottmann (see Johnson, 1980), and may soon be rivalled by the Japanese cities of Tokyo, Osaka and Nagoya, and the urban belt of western Europe stretching from the Ruhr in West Germany through Belgium and The Netherlands, and including north-east France and south-east England.

Despite the continuing increase in the extent of major towns and cities, in many countries the land occupied by the urban population is often less than 5% of the total area (UNESCO, 1979). The concentration of human activities intensifies local competition for all types of resources, among the most vital of which is water. As Schneider *et al.* (1973) have noted, water is both an artery and a vein to urban life. In addition to those uses that are essential for human existence, water is also employed extensively in urban areas for the disposal of wastes. However, for the majority of most urban populations, individual rights and responsibilities in relation to these functions have been delegated to the local community. Of more immediate concern to individuals is the role of water as a nuisance, among the more obvious forms of which are flooding, drainage, erosion and sedimentation. These problems are compounded by the modifications to both the natural environment in general and the landscape in particular that are a consequence of urbanisation. These changes are sufficiently radical to justify separate consideration within the general field of hydrology.

## 1.2. THE DEFINITION OF URBAN HYDROLOGY

Hydrology may be defined as the physical science which treats the waters of the Earth, their occurrence, circulation and distribution, their chemical and physical properties, and their reaction with the environment, including their relation to living things (UNESCO, 1979). These words serve to emphasise two particular aspects of the subject: its interdisciplinary nature, which embraces physical, chemical and biological as well as applied sciences; and its concern with the spatial and temporal distribution and movement of water in all its forms. The latter is implicit in the concept of the hydrological cycle, which illustrates the multifarious paths by which the water precipitated on to the land surface finds its way to the oceans, where evaporation provides the supply of moisture for the renewal of the process.

The hydrological cycle is commonly presented in pictorial form, of which Fig. 1.1, adapted from Todd (1959), provides a typical example.

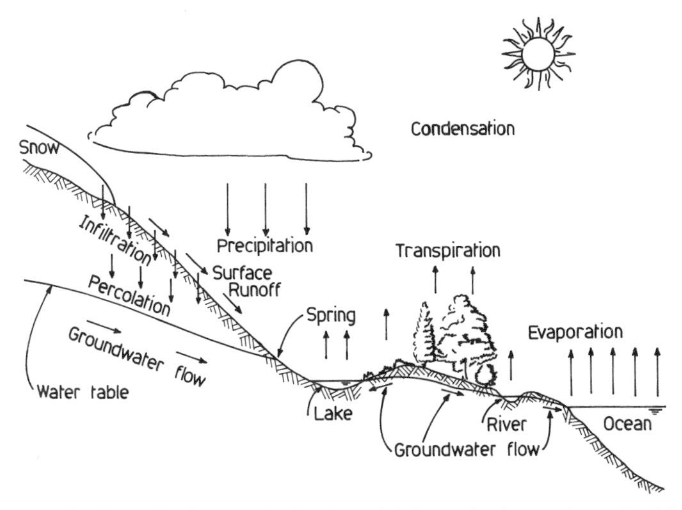


FIG. 1.1. The hydrological cycle in pictorial form (redrawn from Todd, 1959, by permission of John Wiley & Sons, Inc.).

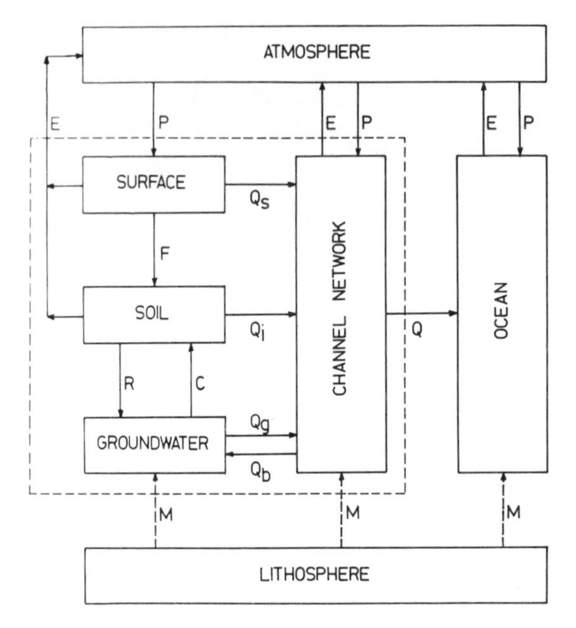


FIG. 1.2. The hydrological cycle in systems notation (modified from Dooge, 1973).

Although Fig. 1.1 is useful in imparting the essential features of a water cycle driven by the excess of incoming over outgoing radiant energy, this representation fails to provide an adequate framework for the study of its component processes. Such a framework can be obtained by adopting the so-called systems notation, in which the paths of water transport link the major sources of moisture storage, as presented by Dooge (1973) in Fig. 1.2.

A closer examination of Fig. 1.2 reveals that hydrologists do not in fact concern themselves with the whole of the hydrological cycle. The oceans are the province of the oceanographer, the atmosphere is studied by the meteorologist, and the lithosphere by the geologist. What remains is commonly referred to as the land phase of the hydrological cycle. This subsystem, whose limits are shown by the broken line in Fig. 1.2, receives an input of precipitation, $P$, and produces outputs in the form of evaporation, $E$, and river flow, $Q$. Further subdivision is possible in order to demarcate the interests of other specialist groups. For example, the soil scientist may confine his interests to the upper soil horizons, which receive water by infiltration, $F$, or capillary rise, $C$, and lose water by evaporation, $E$, deep percolation, $R$, or throughflow, $Q_i$. Nevertheless, despite the improvement in the level of comprehension afforded by Fig. 1.2 over Fig. 1.1, an additional important element is missing—that of the influence of man.

Since time immemorial, man has manipulated his environment, and therefore the land phase of the hydrological cycle, for his own purposes. Wildscape has been cleared for agriculture, forests have been felled, swamps have been drained and, most important of all, towns and cities with all their associated infrastructure have been created in what were once rural areas. Over the last 25 years, increased attention has been devoted to the hydrology of land use changes in general, but only the process of urbanisation has given rise to a new and recognisable branch of the subject—urban hydrology.

Perhaps the most obvious definition of urban hydrology would be the study of the hydrological processes occurring within the urban

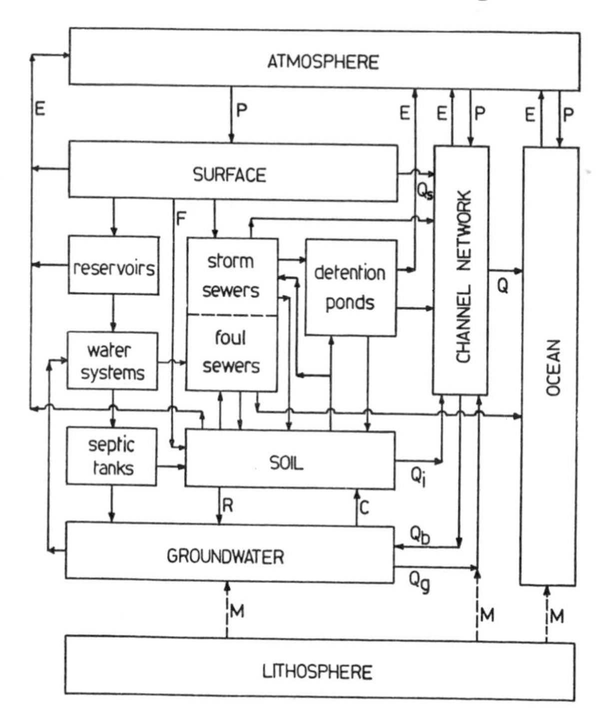


FIG. 1.3. The urban hydrological cycle.

environment. However, further consideration of the hydrological cycle of an urban area, as presented in Fig. 1.3, soon reveals the inadequacy of this simplistic conception. The natural drainage systems are both altered and supplemented by sewerage. The effects of flooding are mitigated by flood alleviation schemes or storage ponds. In the initial stages of urban development, septic tanks are employed for the disposal of domestic wastes. As the urban area grows, foul sewerage systems discharging to sewage treatment works are installed, and the treated effluent is returned to local watercourses or even the ocean. Initially, water supplies are drawn from local surface and groundwater sources at minimum cost, but as the population increases and the demand for water rises, further supplies may only be obtainable from more remote locations. Both waste disposal and water supply therefore extend the influence of the urban area well beyond its immediate boundaries. Urban hydrology may consequently be defined in more appropriate terms as the study of hydrological processes both within and outside the urban environment that are affected by urbanisation.

## 1.3 THE SCOPE OF URBAN HYDROLOGY

Several authors, including Savini and Kammerer (1961), Leopold (1968), Hall (1973) and Cordery (1976), have described the changes in flow regime which occur when an initially rural catchment area is subject to urbanisation. The particular aspects of urbanisation which exert the most obvious influence on hydrological processes are the increase in population density and the increase in building density within the urban area. The consequences of such changes are outlined diagrammatically in Fig. 1.4.

As the population increases, water demand begins to rise. This growth in demand is accelerated as standards of living are raised and compounds the problem of developing adequate water resources—the first of the major hydrological problems.

Once the initial stages of urbanisation have passed and sewerage systems are installed for both domestic and surface water drainage, the amount of waterborne waste increases in response to the growth in population. However, the resultant water quality changes are intimately linked with the consequences of the increase in building density. As the latter rises, the extent of impervious area also increases,

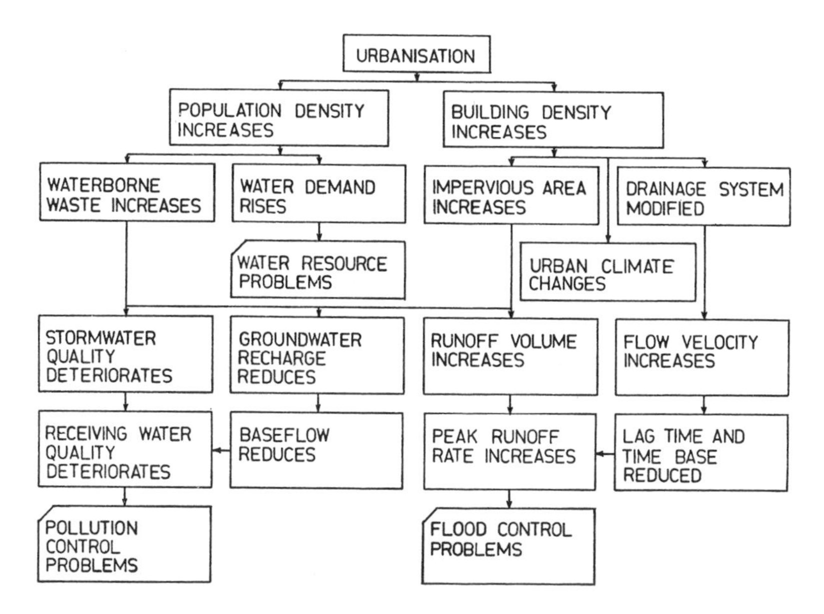


FIG. 1.4. The effects of urbanisation on hydrological processes.

the natural drainage system is modified and the local microclimate changes. Owing to the larger impervious area, a greater proportion of the incident rainfall appears as runoff than was experienced when the catchment was in its rural state. Furthermore, the laying of storm sewers and the realignment and culverting of natural stream channels which takes place during urbanisation result in water being transmitted to the drainage network more rapidly. This increase in flow velocities directly affects the timing of the runoff hydrograph. Since a larger volume of runoff is discharged within a shorter time interval, peak rates of flow inevitably increase, giving rise to the second of the major hydrological problems—flood control.

The inadvertent changes in the microclimate which accompany the growth of urban areas may at first sight appear somewhat irrelevant in comparison to the changes in the hydrological cycle brought about by urbanisation. Nevertheless, further consideration of the available evidence, as presented by Landsberg (1981a, b), for example, shows that, since all aspects of climate are affected to some extent by urban development, some attention should be devoted to the possible consequences of such changes in terms of infrastructure design. For example, in drainage design practice, particular importance is attached to the frequency of heavy rainfalls within predetermined durations. Changes in the relationship between rainfall depth, duration and frequency may therefore alter the degree of protection afforded by engineering works subsequent to their design and construction. Possible allowances for such changes are most conveniently treated as a supplementary aspect of the flood control problem.

As Fig. 1.4 demonstrates, the water quality aspects of the hydrological cycle are affected by both the rise in population and the increase in the extent of the impervious area. Since the volume of runoff becomes larger with the onset of development, the amount of soil moisture recharge is reduced. Consequently, less water is likely to percolate into any aquifer underlying the urban area. Between storm events, the baseflow within the natural drainage system is derived from such subsurface storages. Low flows may therefore be expected to decrease as the urbanisation of an area proceeds. Unfortunately, this decrease occurs simultaneously with the increase in the volume of waterborne wastes referred to above and the deterioration in the quality of stormwater runoff as contaminants are washed from streets, roofs and paved areas. The disposal of both solid and waterborne wastes may also have an adverse effect upon groundwater quality. The degradation of the quality of the flows in both the drainage network serving the urban area and the underlying aquifers gives rise to the third of the major hydrological problems—pollution control.

In summary, the process of urbanisation may be seen to create three major hydrological problems: the provision of water resources for the urban area that are adequate in both quantity and quality; the prevention of flooding within urban areas; and the disposal of waterborne wastes from urban areas without impairing the quality of local watercourses. Of these three problems, that of water supply forms part of the wider subject of water resources development, and is beyond the scope of this text. Nevertheless, two distinct attitudes to the development of water resources for rapidly growing urban areas may be identified in current practice. The principal features of these contrasting viewpoints are therefore reviewed briefly in Section 1.4. The two remaining problems, flood and pollution control, which form the subject of subsequent chapters, are discussed in Section 1.5 prior to outlining the arrangement of contents in Section 1.6.

## 1.4. WATER RESOURCES PROBLEMS OF URBAN AREAS

The supply of clean, pure drinking water and the provision of adequate flows for the disposal of waterborne wastes are essential to those who live in towns and cities. The magnitude of the per capita domestic consumption of water in turn depends largely on the prevailing climate and the prosperity of the inhabitants. In addition, water is required for both industrial processes and recreation and amenity purposes. Production, employment and the pursuit of leisure are therefore intimately related to available water resources, thereby underlining the fundamental role of water services in urban life. Indeed, these services are so basic to the 'quality of life' that their provision rarely provokes the heated debates that often accompany, for example, the planning of airports or motorways. Nevertheless, in many countries the history of water resources planning for towns and cities has consisted of alternating periods of water shortage and renewed investment in which local water supplies have been exploited to satisfy foreseeable local demands. Two typical case studies, relating to Miami and New York City, that illustrate this point have been presented by Schneider and Spieker (1969). However, as urban areas continue to expand, new sources of water become increasingly remote from the population that they serve. Competition for the same resource may then arise between adjacent towns and cities, and the former criterion of satisfying immediate needs at minimum cost clearly becomes redundant.

In the face of competition for the available resources, two approaches to the resolution of the conflict are possible. The first of these is based upon the premise that the ultimate population of an area is limited by the available water supply, which itself is limited to the precipitation which falls within the administrative boundary of the responsible authority. An example of the application of this approach to the Atlantic coastal plain of New Jersey has been discussed by Hordon (1977). In New Jersey, the policy of planning only to the limit of local water supplies appears to have been adopted on largely legislative grounds, the shallow unconfined aquifer within the area being preferred as a resource to the deeper confined formation whose recharge zone lies within the jurisdiction of another authority.

This conservative approach to development inevitably provides a lower limit to the extent of the resources that can be exploited on a truly regional basis. However, the application of a regional strategy requires an appropriate legislative and administrative framework through which its provisions can be implemented. The evolution of water legislation in England and Wales, a synopsis of which has recently been presented by Sewell and Barr (1978), provides a notable example of this type of interaction. The Water Resources Act 1963 provided for the preparation of long-term water resources development plans for England and Wales on both a national and a regional basis. As noted by Rydz (1971), the various options contained within each regional study were formulated without regard to the limited powers of the then river authorities and water undertakings of that region. The required degree of flexibility has since been achieved by the creation of the Regional Water Authorities under the Water Act 1973. Comprehensive discussions of these changes and the advantages which accrue from regional water resources management have been presented by Okun (1977) and Porter (1978).

## 1.5. FLOOD AND POLLUTION CONTROL PROBLEMS OF URBAN AREAS

As noted in Section 1.3, this text is devoted principally to the dual problems of flood control and pollution control. Consideration of the former, as outlined in Fig. 1.4, shows that as the predominant land use of a catchment area changes from rural to urban, the alterations to the form of the landscape and the surface cover are often so radical that the provision of stormwater sewerage becomes essential to protect both property and lives. The design of such sewer networks may be regarded as the 'internal' drainage problem of the urban area. However, once constructed, sewerage systems are instrumental in changing the hydrological regime of the catchment area within which the urban area is located. The increases in runoff volumes, the shorter times for flows to reach their maximum, the increases in peak flow rates and the reduced baseflows are experienced downstream of the urban area, thereby creating an 'external' drainage problem.

Although the internal and external drainage problems are clearly intimately related, they have traditionally been considered in isolation, the former being the concern of the municipal engineer and the latter that of the land drainage or river engineer. Indeed, the design procedures applied by both groups have been broadly similar, as illustrated by the flowchart shown in Fig. 1.5.

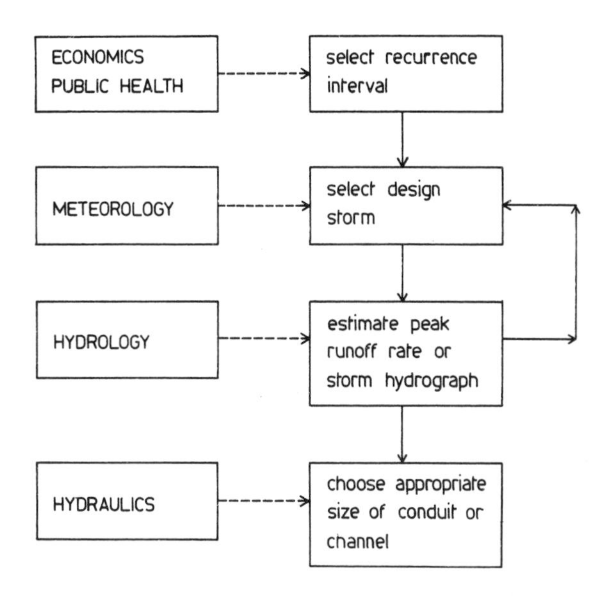


FIG. 1.5. Design procedure for drainage problems.

Four distinct steps may be identified in the design procedure. Firstly, the degree of protection to be provided by the works, i.e. the tolerable frequency of flooding, must be identified. With regard to internal drainage problems, flooding is defined by the conditions under which the sewers have insufficient capacity to carry away the peak flow rate. Although reverse flow may occur at road gulleys and manhole covers may lift, the drainage network rarely fails in the structural sense of the term. However, in treating external drainage problems, the structural integrity of the works forms a fundamental design consideration. In either case, the design frequency, or its reciprocal, the return period, should ideally be based upon an economic analysis, in which the benefits of the scheme in terms of the damage costs avoided are balanced against construction costs. Unfortunately although tangible benefits, such as property damage, are readily quantified, the intangible benefits, such as general inconvenience, transport delays and injuries to health, are much more difficult to assess. Evaluation of both types of benefit can therefore involve a considerable expenditure of time and effort which may only be justified for works involving large capital investments, such as major urban flood protection schemes. For lesser projects, design standards have been evolved whose purpose is to ensure that the works are capable of withstanding a wide range of stresses during their working life. Such standards allow implicitly for both the value of the property that the drainage scheme will protect and, if the sewers carry domestic waste as well as stormwater, the possible danger to public health in system overflows.

Once the tolerable flooding frequency has been determined, the second step in the procedure of Fig. 1.5 involves the acquisition of the appropriate rainfall information for design purposes. The 'design storm' may consist of an average rainfall rate corresponding to a specified storm duration, or a storm profile showing the variation of rainfall intensity with time throughout the event. In any design application, the type of meteorological information depends primarily upon the method of flood estimation that is adopted in the third step of the procedure, thereby giving rise to the feedback of information shown in Fig. 1.5.

Having constructed the design storm, the third step in the procedure consists of the transformation of the rainfall into runoff. There are three important aspects of this transformation:

(1) the proportion of the total volume of rainfall that appears as surface runoff;
(2) the manner in which the runoff volume is distributed in time, i.e. the shape of the flood hydrograph; and
(3) the relationship between the frequency of the design rainfall and the frequency of the resultant peak rate of flow.

The final step in the procedure involves the determination of the size of channel or conduit required to carry away the estimated peak flow rate. For sewer systems, design tables giving the capacities and flow velocities for pipes of a given diameter and roughness laid at a specified gradient are available for use in unsurcharged flows (see Hydraulics Research Station, 1977). With regard to external drainage problems, structural measures, such as flood alleviation channels and storage ponds, have invariably been chosen as the most appropriate devices to mitigate the effects of urbanisation. However, with enlarged channels, care is necessary to ensure that the original flooding problem is not simply passed further downstream. In addition, where several flood storage ponds are constructed within a catchment area, the designer must ensure that their operation does not aggravate rather than lessen the amount of flooding. Recent experience in the United

States has tended to indicate that engineering works alone may not be completely successful in reducing flood losses. A variety of non-structural measures, including land use controls, floodproofing and insurance programmes, have been instituted in an effort to avoid heavy expenditure on projects which fail to achieve their objectives. White (1975) has presented a comprehensive review of the factors which enter into the development of optimal policies for the reduction of flood damages.

Although structural and non-structural flood control measures are essential to the management of water in a quantitative sense, Fig. 1.4 shows that such strategies cannot be divorced entirely from water quality considerations. Until the early 1960s, water quality control strategies were directed primarily at improvements in treating domestic and industrial wastes. The higher quality of effluent achieved at many sewage treatment works as a result of these efforts has served only to draw greater attention to the degradation of receiving watercourses that can be caused by diffused or non-point sources of pollutants, such as urban storm runoff, and uncontrolled point sources, such as storm-water overflows. By their very nature, non-point pollution problems are not amenable to technical solutions involving collection and treatment practices such as those applied to domestic wastes. Planning measures, such as the control of land use and the preservation of natural vegetative cover, appear to hold greater prospects for success, but depend upon a clear understanding of the mechanisms which govern temporal and spatial variations in water quality. An appreciation of this variability is also fundamental to the construction of mathematical models which can be used in the formulation of water pollution control strategies.

## 1.6. ARRANGEMENT OF CONTENTS

The flowchart illustrating the design procedure that is followed in treating both internal and external drainage problems presented in Fig. 1.5 shows the diversity of information that must be called upon during its execution. Although water quality considerations are not explicitly included in Fig. 1.5, a designer must also possess an awareness of the implications of his proposals in this context. For convenience in presentation of this background knowledge, the following text has been divided into four parts.

Part I of the text contains two chapters which deal with meteorological aspects of drainage design. The principal features of the urban climate are described in Chapter 2, which lays particular emphasis on the implications of rainfall anomalies in urban areas. In Chapter 3, attention is turned to the rainfall data that are required for flood estimation and drainage design. In addition to discussing point rainfall statistics, i.e. the information that can be derived from the records of a single raingauge, the description of temporal and spatial variations in rainfall depths is treated along with that of storm movement.

Part II of the text consists of four chapters that are devoted to flood hydrology. Chapter 4 provides a general introduction to problems of design flood estimation, and outlines a framework within which different methods of approach may be compared and their suitability for application to particular types of problem assessed. Two broad categories of approaches to design flood estimation may be identified: statistical methods, which depend upon the fitting of probability distributions to series of recorded peak rates of flow; and deterministic methods, which rely on the identification of a relationship between rainfall and streamflow that can be employed to synthesise frequency information on flows from the available rainfall statistics. The statistical methods are outlined in Chapter 5. Deterministic methods may be considered to involve the description of two distinct but related processes: the transformation of rainfall into runoff; and the routing of the runoff through the complex of conduits and channels which form the drainage network of the catchment. Rainfall–runoff relationships are discussed in Chapter 6, and flood routing in Chapter 7.

The principal objective of Parts I and II is to provide introductory material for the discussion of the hydrological and water quality problems that are addressed in Parts III and IV respectively. In Part III, Chapter 8 deals with the external drainage problem, i.e. the description of the changes in flow regime caused by urbanisation, and Chapter 9 outlines approaches to the internal drainage problem posed by stormwater drainage design. The discussion of water quality problems in Part IV begins with a review of the water quality changes caused by urbanisation in Chapter 10. Chapter 11 then outlines the range of modelling techniques that has been applied to describe the variations in water quality experienced in urban areas. Finally, Chapter 12 draws together the quantity and quality aspects by considering the overall implications of infrastructure development in urban areas, ranging from initial construction to the implementation of flood control measures, such as channelisation and the building of flood storage ponds, as well as the potential nuisance created by the de-icing of roads.

## REFERENCES

CORDERY, I. (1976) Some effects of urbanisation on streams. *Civ. Engng. Trans., Instn. Engrs. Austr.*, **CE18**(1), 7–11.

DOOGE, J. C. I. (1973) Linear theory of hydrologic systems. US Dept. Agric., Agric. Res. Serv., Tech. Bull. 1468, 327 pp.

HALL, M. J. (1973) The hydrological consequences of urbanisation: an introductory note. In Construction Industry Research and Information Association, *Proc. Res. Colloquium on Rainfall, Runoff and Surface Water Drainage of Urban Catchments*, Bristol, paper 10.

HORDON, R. M. (1977) Water supply as a limiting factor in developing communities. Int. Assoc. Hydrol. Sci., Publ. No. 123, pp. 520–5.

HYDRAULICS RESEARCH STATION (1977) *Tables for the hydraulic design of pipes, metric edition* (HMSO, London) 144 pp.

JOHNSON, J. H. (1980) *Urbanisation* (Macmillan, London) 60 pp.

LANDSBERG, H. E. (1981a) City climate. Ch. 3 of Landsberg, H. E. (ed.), *General climatology* 3, *World survey of climatology*, Vol. 3 (Elsevier, Amsterdam) pp. 299–334.

LANDSBERG, H. E. (1981b) *The urban climate* (Academic Press, New York) 275 pp.

LEOPOLD, L. B. (1968) Hydrology for urban land planning—a guidebook on the hydrologic effects of urban land use. US Geol. Survey, Circ. 554, 18 pp.

OKUN, D. A. (1977) *Regionalisation in water management: a revolution in England and Wales* (Applied Science Publishers, London) 377 pp.

PORTER, E. (1978) *Water management in England and Wales* (Cambridge Univ. Press) 178 pp.

RYDZ, B. (1971) Regional water resources analysis. *Proc. Instn. Civ. Engrs.*, **49,** 129–43.

SAVINI, J. and KAMMERER, J. C. (1961) Urban growth and the water regimen. US Geol. Survey, Water-Supply Paper 1591-A, 43 pp.

SCHNEIDER, W. J. and SPIEKER, A. M. (1969) Water for the cities—the outlook. US Geol. Survey, Circ. 601-A, 6 pp.

SCHNEIDER, W. J., RICKERT, D. A. and SPIEKER, A. M. (1973) Role of water in urban planning and management. US Geol. Survey, Circ. 601-H, 10 pp.

SEWELL, W. R. D. and BARR, L. R. (1978) Water administration in England and Wales: impacts of regionalisation, *Wat. Resour. Bull.*, **14,** 337–49.

TODD, D. K. (1959) *Groundwater hydrology*, 1st Edn (Wiley, New York) 336 pp.

UNITED NATIONS EDUCATIONAL, SCIENTIFIC AND CULTURAL ORGANISATION (1979) Impact of urbanisation and industrialisation on water resources planning and management. UNESCO Studies and Repts. in Hydrology, No. 26, 111 pp.

WHITE, G. F. (1975) Flood damage prevention policies. *Nature and Resources*, **11**(1), 2–7.

# Part I
# METEOROLOGY

# 2

# Urban Climate

## 2.1. INTRODUCTION

Ever since people have congregated together in towns and cities because of either their own inherent gregariousness or a mutual need for common defence, they have attempted to change their immediate surroundings. The extent to which the change is felt to be necessary depends primarily upon the need for shelter in the prevailing climate. The inevitable by-product of any such attempt to create a pocket of controlled environment is an inadvertent and often subtle alteration to the local microclimate.

All climatic elements are modified to a certain extent in any urban conurbation. In creating his towns and cities, man has radically altered the surface of the ground. As a result, the radiation balance of the area is modified and the change in the aerodynamic roughness affects air motion. Heat is added to the atmosphere by both domestic and industrial heating processes and by motor vehicles. In addition, combustion processes add water vapour to the atmosphere along with a wide variety of chemicals. However, owing to the complex nature of the physical processes which determine the climate of a given locality, the contrast between towns and cities and their surrounding rural areas is difficult to evaluate. This problem arises from the tendency for settlements to grow where there are features in the local geography, such as a natural harbour, the confluence of two river valleys, a lake or an estuary, that could be employed to advantage for either trade or defensive purposes. Unfortunately, such features also exhibit local microclimates that differ from that of the surrounding area.

A framework for examining the differences in climate between urban areas and their rural surroundings has been provided by Lowry

(1977) in terms of the general equation

$$M(i, t, x) = C(i, x) + L(i, x) + E(i, t, x) \tag{2.1}$$

where $M$ is the observation of a climate variable in synoptic weather type $i$ at time $t$ and site $x$; $C$ is the background climate, i.e. that of a flat plain with no landscape or urban effects, and is assumed to be time-invariant; $L$ is the departure from $C$ that is caused by topography and geographical features, and is also assumed to be time-invariant; and $E$ is the departure from $C$ caused by urbanisation. Contemporaneous urban–rural differences in the climatic variable may then be evaluated using data from an urban ($x = u$) and a rural ($x = r$) site. Assuming that $E(i, t, r)$ is zero at the rural site, the effect of urbanisation is given by

$$\begin{aligned} \Delta(i, t, u - r) &= M(i, t, u) - M(i, t, r) \\ &= C(i, u) - C(i, r) + L(i, u) - L(i, r) + E(i, t, u) \end{aligned} \tag{2.2}$$

This equation shows that $\Delta$ is equal to $E(i, t, u)$ only if landscape effects can be ignored and the two sites lie on the same isopleth of the 'flat plain' climate. Other measures of urbanisation, such as contemporaneous upwind–downwind differences, contemporaneous urban–regional ratios, time trends in the differences and ratios and contemporaneous weekday–weekend differences, can be shown to be similarly confounded by the influences of regional climate and topography. As Oke (1979) has noted, the only measure that can be used with any confidence is that between urban and pre-urban conditions stratified by synoptic weather type.

One of the first quantitative essays on urban climate, written on that of London by the pioneer nineteenth-century meteorologist Luke Howard, was first published between 1818 and 1820, with a revised edition appearing in 1833 (see Blench, 1963). More than a century later, comprehensive surveys of the progress that has been made in understanding urban climates, based largely on studies in Europe and North America, have been provided by Landsberg (1956, 1961, 1981a, b) and Oke (1974, 1979). Table 2.1, quoted from Landsberg (1981b), illustrates the order of magnitude of the long-term changes in climate that may be produced by towns and cities. However, in particular localities and over different time scales, individual climatic elements may show markedly different variations, and so the contrasts between urban and rural areas are better demonstrated with reference to specific sites. The discussions contained in the following sections

**Table 2.1. Climatic alterations produced by cities**
(after Landsberg, 1981b, by permission of Academic Press, Inc.)

| | *Element* | *Comparison with rural area* |
|---|---|---|
| Contaminants: | Condensation nuclei | 10 times more |
| | Particulates | 10 times more |
| | Gaseous admixtures | 5–25 times more |
| Radiation: | Total on horizontal surface | 0–20% less |
| | Ultraviolet—winter | 30% less |
| | —summer | 5% less |
| | Sunshine duration | 5–15% less |
| Cloudiness: | Clouds | 5–10% more |
| | Fog—winter | 100% more |
| | —summer | 30% more |
| Precipitation: | Amounts | 5–15% more |
| | Days with <5 mm | 10% more |
| | Snowfall—inner city | 5–10% less |
| | —lee of city | 10% more |
| | Thunderstorms | 10–15% more |
| Temperature: | Annual mean | 0·5–3·0 deg C more |
| | Winter minima (avge) | 1–2 deg C more |
| | Summer maxima | 1–3 deg C more |
| | Heating degree days | 10% less |
| Relative humidity: | Annual mean | 6% less |
| | Winter | 2% less |
| | Summer | 8% less |
| Wind speed: | Annual mean | 20–30% less |
| | Extreme gusts | 10–20% less |
| | Calm | 5–20% more |

therefore draw liberally upon the results presented by Chandler (1965) in his study of the climate of London.

The Greater London area, which now has a population of over 7 million, covers approximately 2000 km$^2$ of south-east England lying between the chalk outcrops of the North Downs to the south and the Chiltern Hills to the north-west. London itself has grown outwards from two nuclei located in the City of London and Westminster. Surrounding these nuclei are three distinct bands of development, each dating from a different period. The first, containing the West End and the City, consists predominantly of closely spaced modern and Victorian buildings, but is enlivened by large open spaces, such as Hyde Park, Green Park and St James's Park. The second band, which developed between 1750 and 1914, encompasses a number of high-density residential and industrial areas, with dockland to the east and more open spaces, such as Regent's Park, Richmond Park, Kew Gardens and Hampstead Heath. The third band contains mainly post-1914 development, with two-storey detached and semi-detached housing and some industry along the main roads.

Despite the complexity of this development over the centuries, during which the expanding urban area has engulfed what were once outlying villages, the available climatological observations are sufficient to illustrate changes in the microclimate of London that are characteristic of many other major modern cities. These changes are reviewed in Sections 2.2 to 2.4 under the general headings of heat production, atmospheric composition, and surface geometry and roughness. The chapter concludes in Section 2.5 with a discussion of the impact of these climatic alterations on the planning of urban infrastructure in general and water supply and drainage systems in particular.

## 2.2. HEAT PRODUCTION IN AN URBAN AREA

Owing to the higher thermal conductivity and greater heat capacity of the fabric of buildings compared with those of vegetated areas, the thermal properties of urban areas contrast strongly with those of their rural environs. The temperature changes that result from these differing thermal characteristics are typically illustrated by the average annual temperatures for the London area compiled by Chandler (1965) and summarised in Table 2.2. These data show that the mean, maximum and minimum temperatures in central districts are all higher than those in the suburbs, margins and surrounding country. The smallest differences are exhibited by the maximum and the largest by the minimum temperatures, although the magnitude of the changes is partly influenced by station altitude within the margins and suburbs. This positive temperature anomaly in favour of the central districts is referred to as the 'urban heat island', a term which according to Landsberg (1981b) was first used in the English-language meteorological literature by Manley (1958).

**Table 2.2. Average annual temperatures (°C) for the London area, 1931–1960**
(after Chandler, 1965)

| | *Height (m)* | *Average annual temperature (°C)* | | |
|---|---|---|---|---|
| | | *Maximum* | *Minimum* | *Mean* |
| Surrounding country | 87·5 | 13·7 | 5·5 | 9·6 |
| Margins—high level | 144·2 | 13·4 | 6·2 | 9·8 |
| Suburbs—high level | 137·2 | 13·4 | 5·9 | 9·7 |
| Suburbs—low level | 61·9 | 14·2 | 6·4 | 10·3 |
| Central districts | 26·5 | 14·6 | 7·4 | 11·0 |

Oke (1982) has observed that urban heat islands have been well described but poorly understood. According to the same author (Oke, 1979, 1982), a better appreciation of the nature of the temperature anomaly can be achieved if the layers of the atmosphere below and above roof level, referred to as the urban canopy layer and the urban boundary layer respectively, are considered separately. Table 2.3, taken from Oke (1982), lists the components of the energy balance that contribute towards the formation of the heat island and the features of urbanisation which cause them to change. This table shows that the radiation balance of the urban canopy layer is particularly dependent upon the depth and orientation of the 'canyons' created by the street layout. The resultant complexities of urban temperature fields have been revealed by measurements taken from specially instrumented vehicles in London by Chandler (1965) and in Dallas and Fort Worth by Ludwig (1970), among others. Ludwig (1970) showed that under daytime summer conditions, a 'warm ring' may develop within an urban area separating the city centre from the even cooler outskirts. This particular form of heat island was attributed largely to the height and spacing of the buildings. With the tower blocks which distinguish the central business districts of most modern cities, daytime incoming radiation is absorbed at higher levels and the street is effectively shielded. In contrast, areas with closely spaced buildings of a lesser height tend to be warmer, probably because multiple reflections lead to a greater absorption of insolation.

Urban heat islands have been found to exhibit marked temporal as well as spatial variations. For example, Chandler (1965) has shown

**Table 2.3. Possible causes of the urban heat island: an unranked listing** (after Oke, 1982, by permission)

| *Altered energy balance terms leading to positive thermal anomaly* | *Features of urbanisation underlying energy balance changes* |
|---|---|
| A. Urban canopy layer: | |
| 1. Increased absorption of short-wave radiation | Canyon geometry—increased surface area and multiple reflection |
| 2. Increased long-wave radiation from the sky | Air pollution—greater absorption and re-emission |
| 3. Decreased long-wave radiation loss | Canyon geometry—reduction of sky view factor |
| 4. Anthropogenic heat source | Building and traffic heat losses |
| 5. Increased sensible heat storage | Construction materials—increased thermal admittance |
| 6. Decreased evapotranspiration | Construction materials—increased 'waterproofing' |
| 7. Decreased total turbulent heat transport | Canyon geometry—reduction of wind speed |
| B. Urban boundary layer: | |
| 1. Increased absorption of short-wave radiation | Air pollution—increased aerosol absorption |
| 2. Anthropogenic heat source | Chimney and stack heat losses |
| 3. Increased sensible heat input—entrainment from below | Canopy heat island—increased heat flux from canopy layer and roofs |
| 4. Increased sensible heat input—entrainment from above | Heat island and roughness—increased turbulent entrainment |

that the intensity of London's heat island displays a marked seasonality, with the largest temperature anomalies in central districts occurring during the summer months of June and July. The same seasonality was not evident in the suburbs, an effect attributable to the lower building densities, and the more frequent fogs which reduce radiation heat losses.

When the behaviour of the urban heat island is examined on a diurnal basis, the largest positive anomalies are found to develop 1–2 h before dawn and the smallest in mid-afternoon. There is generally a delay of some 1–2 h between the maximum temperatures at urban and rural sites. The minimum temperatures are achieved later in the urban area and the temperature range is generally smaller. These features may be found in the mean diurnal temperature variations for Frankfurt

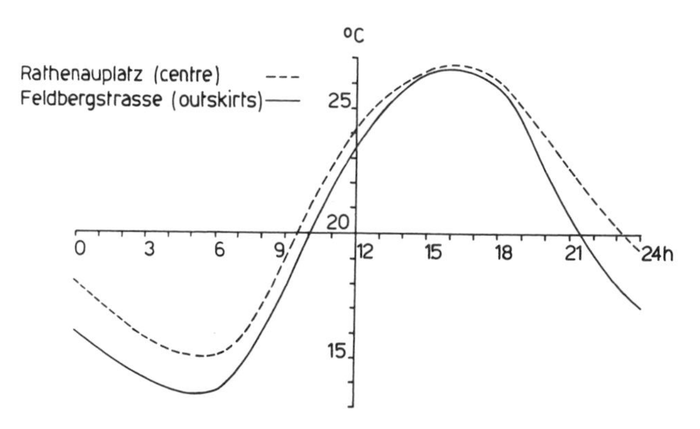


FIG. 2.1. Mean diurnal summer temperature variation for Frankfurt, 1959–1963 (redrawn from Georgii, 1970, by permission of the World Meteorological Organisation).

quoted by Georgii (1970) and reproduced in Fig. 2.1. During the winter months the diurnal temperature variation is generally similar, although its amplitude may be smaller, as Mitchell (1961) has shown for Vienna. In addition, during the summer, temperatures in the country may rise more rapidly than in the urban area. This 'urban cold island' may occur partly because of the differences between the thermal properties of buildings and vegetated soils and partly because of the attenuation of sunlight within the urban atmosphere. The occurrence of a cold island within the city of Birmingham was found by Unwin (1980) to be associated with disturbed types of airflow, which suggests that vertical mixing may also be an important factor in their development.

According to Chandler (1965), London's urban heat island is most marked when skies are clear and wind speeds do not exceed 5–6 m/s. Similarly, Oke and Hannell (1970) have reported that the heat island of Hamilton, a town of some 300 000 inhabitants in Ontario, is reduced to insignificant proportions when wind speeds exceed 6–8 m/s.

## 2.3. CHANGES IN THE COMPOSITION OF THE ATMOSPHERE

As Table 2.1 indicates, the urban atmosphere is characterised by a marked abundance of dust particles along with sulphur dioxide and

**Table 2.4. Averages of bright sunshine for the London area, 1921–1950**
(after Chandler, 1965)

| | *Average (h/day)* | *Loss (min/day)* |
|---|---|---|
| Surrounding country | 4·33 | — |
| Outer suburbs | 4·07 | 16 |
| Inner high-level suburbs | 4·03 | 18 |
| Inner low-level suburbs | 3·95 | 25 |
| Central districts | 3·60 | 44 |

other gases. These contaminants not only reduce the clarity of the atmosphere, thereby decreasing the amount of incoming radiation and sunshine, but also provide an excess of condensation nuclei that may change the nature of city fogs and affect the characteristics of precipitation.

Chandler (1965) has shown that the effect of smoke and gases on the averages of bright sunshine in the London area during the period 1921–1950 was particularly marked. On average, the outer suburbs were found to lose some 16 min, the inner suburbs up to 25 min and the city centre about 44 min of bright sunshine per day (see Table 2.4). However, the enactment of the 'clean air' legislation during the 1950s has served to decrease substantially the total smoke emissions in the London area and consequently much smaller losses of mean daily sunshine are now apparent (see Jenkins, 1969). Nevertheless, in spite of this legislation, adverse meteorological conditions may still produce medically undesirable levels of air pollution, as Elsom (1979) has shown in his discussion of such an episode in the Greater Manchester area during the winter of 1970–71.

Owing to the annual variation in the elevation of the sun, urban sunshine losses have a strong seasonal character. During the winter months when the sun's elevation is low, insolation is reduced by absorption after scattering. The centre therefore receives less than half, and the suburbs about three-quarters, of rural sunshine totals. In contrast, during the summer months the centre receives more than 90% of the rural sunshine totals and the difference between the centre and the suburbs is small. Since the high-level suburbs receive more solar radiation and do not experience fogs as frequently as the low-

**Table 2.5. Estimated fog frequencies (h/year) for the London area, 1947–1956**
(after Shellard, 1959, by permission of Her Majesty's Stationery Office)

| | *Visibility (m)* | | | |
|---|---|---|---|---|
| | *<40* | *<200* | *<400* | *<1000* |
| Kingsway | 19 | 126 | 230 | 940 |
| Kew | 79 | 213 | 365 | 633 |
| London Airport | 46 | 209 | 304 | 562 |
| South-east England[a] | 20 | 177 | 261 | 494 |

[a] Average of 7 stations.

level suburbs, the magnitude of the loss in mean daily sunshine is also affected by elevation, as shown in Table 2.4.

The temporal and spatial variations in visibility within an urban area are governed not only by the degree of air pollution but also by lateral and vertical temperature distributions, humidity and wind. Chandler (1965) has noted that the nature and frequency of London's fogs are more characteristic than their density, and this observation is fully supported by the fog frequency data for London compiled by Shellard (1959) which are summarised in Table 2.5. This table shows that, although occasions with visibility less than one kilometre (the internationally accepted definition of a fog) are most frequent at the central site (Kingsway), denser fogs are more common at suburban stations (Kew and London Airport). In general, the fog does not become as dense in the city centre as in the rural surroundings because of the higher temperatures and lower wind speeds and humidities in the built-up area.

Although the condensation nuclei produced by cities and the thermal and mechanical turbulence that they are capable of generating might be expected to affect local cloud cover and precipitation, the available evidence has often tended to be somewhat conflicting. Inevitably, studies designed to evaluate the differences in rainfall patterns between city and rural sites suffer from the inherent variability of the element, contrasts in the exposure of instruments, and the influence of regional climate and local topography, as already discussed in Section 2.1. Several of the early studies in the United States were based upon concurrent time series of rainfall data at a city station and an outlying airport, but such observations became available largely through the growth of internal airline services and not from any scientific initiative. In these circumstances, many of the results obtained have remained open to some doubt. Perhaps the most controversial findings on urban–rural differences in precipitation were those reported by Changnon (1968) for the town of La Porte, which is situated in Indiana some 18 km south-east of the southern end of Lake Michigan. According to Changnon, La Porte received 31% more precipitation, 38% more thunderstorms and 246% more hail-days than its environs during the period 1951–1965. However, these results, which were ascribed to the presence of the Chicago iron and steel industrial complex about 50 km to the west, were subsequently attributed to observer error by Holzman and Thom (1970).

One of the first modern studies of urban effects upon precipitation appears to have been that by Ashworth (1929), who demonstrated that for the industrial town of Rochdale in Lancashire during the period 1918–1927, Sunday was the day of the week with the least rain. Moreover, weekdays had 14% more hours of rain between 06.00 h and 18.00 h than Sundays. More recently, Dettwiller (1970) has shown that the intensity and frequency of rain in Paris are higher on weekdays than on Saturdays and Sundays. Furthermore, the differences were found to be more significant during 1960–1967 than during 1953–1959, a change which was attributed to the steady increase in industrial growth over the period of record.

Similar changes in precipitation patterns have also been detected for non-industrial areas. For example, the university–residential community of Champaign–Urbana, Illinois, which is situated some 210 km south of Chicago, has been the subject of a most comprehensive investigation by Changnon (1961, 1969). This town is located in a flat, featureless glacial plain with no significant changes in relief for 160 km, no industry for 65 km and no large water bodies. Using data for the period 1950–1959, Changnon (1961) was able to show that the average annual precipitation was 12% higher on the lee side of the town than in the surrounding rural area. On a seasonal basis, the largest excesses of urban over rural precipitation occurred in winter (16%), with autumn (14%), spring (13%) and summer (7%) next in order of magnitude. In addition, there were 13% more days with rainfalls of between 0·1 and 6·0 mm in the urban area than at the airport 5·6 km to the south, but no significant difference on days with precipitation exceeding 6·0 mm. The urban area also recorded three more thunderstorm-days per year than the airport.

In order to investigate the possibility that these urban–rural differences were attributable to the natural variability of rainfall, the results obtained were compared with those from a raingauge network in a rural area some 90 km to the north-west. The preliminary conclusion that the enhanced urban area precipitation could have been caused by rainfall variability (Changnon, 1961) was later reversed on the basis of a supplementary study of data for a 19-year period (Changnon, 1969). The latter report also provided evidence for an urban effect on precipitation in four large American cities: Chicago, St Louis, Washington and New York.

With regard to the influence of London's urban area on precipitation, Chandler (1965) has drawn attention to the range of some 230–255 mm in average annual precipitation over the Greater London area, which was thought to be governed largely by regional trends and orographic influences. However, Atkinson (1968, 1969, 1970) has argued that, since urban areas act as both mechanical obstacles and thermal sources, their influence is liable to be reflected more in convective motions such as thunder rainfall. Figure 2.2, taken from Atkinson (1968), shows the number of days when thunder was recorded overhead at 608 raingauges in the south-east of England, 1951–1960. Despite the possibility of errors occurring, either through the slackness of observers or gaps in the raingauge network greater

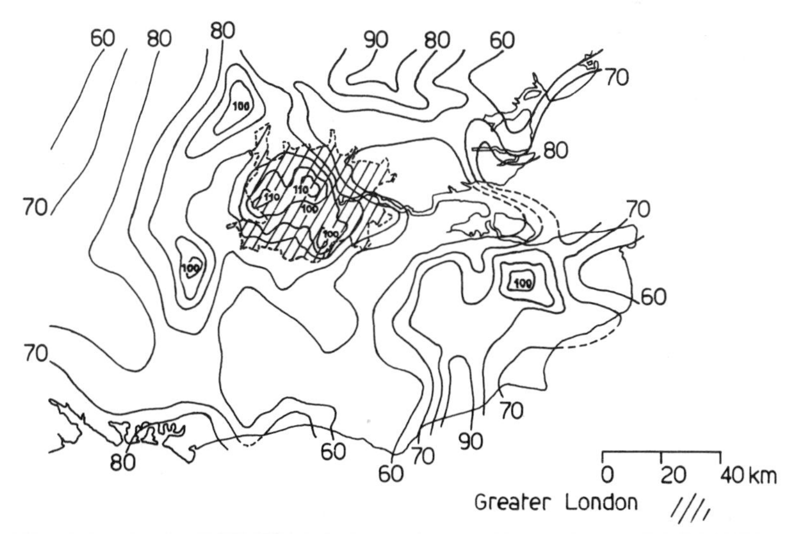


FIG. 2.2. Number of days when thunder was recorded overhead in south-east England, 1951–1960 (redrawn from Atkinson, 1968, by permission).

than the audible range of thunder, the effect of the London conurbation is obvious. Changnon (1980) has also shown that for seven American cities with populations exceeding one million, the maximum increase in thunder-days relative to their surroundings increases by about 8% for every additional one million inhabitants.

Apart from those studies in which thunder rainfall has been considered, few investigations of the urban effect upon precipitation have examined the possibility that changes may take place in the frequency of heavy rainfalls within short durations. This dearth of information is all the more serious in view of its importance in urban drainage design. Among the few results that are available, Brater (1968) drew attention to the tendency for rainfall depths corresponding to return periods exceeding 10 years to be some 20% larger in Detroit, Michigan, than in the surrounding rural area between June and October; from November to May no differences were discernible. Corroboration of these findings only became available with the publication of the results from METROMEX (Metropolitan Meteorological Experiment), a 7-year research programme in St Louis, Missouri, involving the deployment of 225 autographic raingauges and weather radar, which must rank as one of the most comprehensive studies on the urban effects on precipitation ever mounted.

As summarised by Changnon *et al.* (1977), METROMEX identified a region on the lee side of St Louis in which a positive precipitation anomaly was maintained consistently during the summer months throughout a 5-year period, 1971–1975. This anomaly was most pronounced in June, and was more obvious in below-normal rainfall months. Analysis of individual storms revealed that the anomaly was associated with a relatively small proportion of the total number of events which occurred in organised storm systems producing widespread heavy rain. The excess rainfall was largely caused by storms exceeding a depth of 25 mm and having a duration of 3 h or less. Consequently, the frequency distributions of heavy rainfalls for durations between 5 min and 2 h were found to vary significantly between urban, suburban and rural areas (see also Huff, 1977). Should this result prove to be more widely applicable, the design standards achieved on many existing sewerage systems, dimensioned on the basis of country-wide rainfall statistics (see Chapter 3), may be subject to substantial revision.

Finally, since snow tends to fall on windy days when the temperature difference between a city and its surroundings is likely to be small, the frequency of this form of precipitation is unlikely to be affected by the presence of an urban area. Nevertheless, in the absence of orographic influences, owing to the presence of the urban heat island, snow melts more quickly in the city centre than in the suburbs, and more quickly in the suburbs than in the surrounding country.

## 2.4. CHANGES IN SURFACE GEOMETRY AND ROUGHNESS

The mechanical turbulence induced by buildings has already been referred to in discussing the urban effect on precipitation in Section 2.3. When coupled with the thermal turbulence which results from urban heat production, the structure of a city may be seen to exert a considerable influence on air movement. The localised wind effects which accrue from the manner in which the streets are laid out relative to the prevailing wind direction and the juxtaposition of buildings of different height are now widely appreciated (see Penwarden, 1973). However, in order to understand the more general effects on air movement of the geometry and aerodynamic roughness of an urban area, attention must first be paid to the expected seasonal and diurnal variations in wind speeds.

Figure 2.3(a) shows the average diurnal variation for 1932–1956, and Fig. 2.3(b) the average annual variation for 1927–1956 in the wind speed at Kew Observatory in the south-west suburbs of London, as presented by Chandler (1965). Figure 2.3(a) shows that wind speeds are stronger on average by day than by night. This daytime maximum may be attributed to surface heating causing convection which intensifies turbulence, bringing down faster-moving air from above and compensating for the mechanical effects of surface roughness. By night the air is more stable, surface roughness predominates and a marked gradient of wind speed with height develops. Figure 2.3(b) shows that on an annual basis, wind speeds are stronger on average during the winter half-year.

Given the background provided by Fig. 2.3, the urban effect on wind speed may be illustrated by comparing the observations from London Airport with those from Kingsway in Central London. These data, which were analysed by Chandler (1965), are summarised in Table 2.6. This table shows that by night urban wind speeds may exceed rural wind speeds throughout the year. This effect may be explained by the

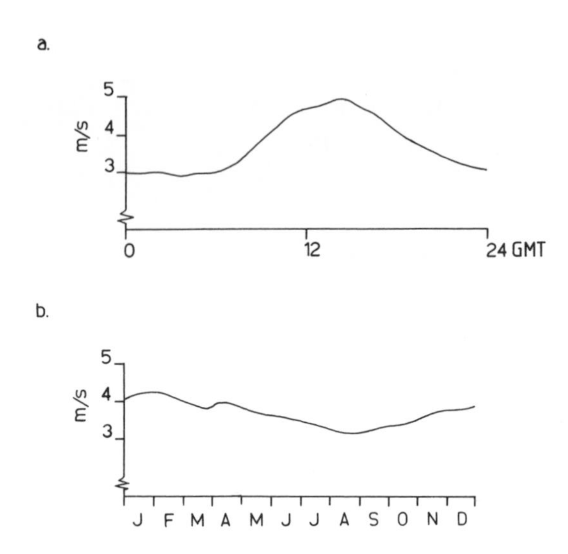


FIG. 2.3. (a) Average diurnal variation, 1932–1956, and (b) average annual variation, 1927–1956, in wind speed at Kew (redrawn from Chandler, 1965).

greater turbulence within the city bringing down faster-moving air to the surface at a time when rural winds are light and rural air stable. By day, when winds are stronger, the air is less stable and rural wind speeds are more uniform with height through turbulent mixing. Within the urban area, however, surface roughness effects predominate and so the wind speeds are lower than in the surrounding rural area. In general, strong winds tend to be reduced in an urban area, whereas

**Table 2.6. Average wind speed (m/s) at London Airport, and its excess over that at Kingsway, 1961–1962**
(after Chandler, 1965)

| | *01.00 GMT* | | *13.00 GMT* | |
|---|---|---|---|---|
| | *Mean* | *Excess* | *Mean* | *Excess* |
| Dec.–Feb. | 2·5 | −0·4 | 3·1 | 0·4 |
| March–May | 2·2 | −0·1 | 3·1 | 1·2 |
| June–Aug. | 2·0 | −0·6 | 2·7 | 0·7 |
| Sept.–Nov. | 2·1 | −0·2 | 2·6 | 0·6 |
| Year | 2·2 | −0·3 | 2·9 | 0·7 |

light winds are increased. There are therefore fewer calms within a city than on its outskirts.

Owing to the reduction in vegetal cover which accompanies urbanisation and the efficiency of man-made drainage systems, which are designed to remove standing water from the surface as quickly as possible, local evaporation is decreased substantially. The absolute humidity within an urban area might therefore be expected to be lower than that of its environs. Exceptionally, in semi-arid areas, excessive irrigation of lawns and parklands may serve to increase the absolute humidity. However, since the water vapour content of the air is generally derived from well outside the built-up area, higher and lower vapour pressures could probably be expected with almost equal frequency. In contrast, the relative humidity may be markedly reduced in the centre of a city owing to the presence of the urban heat island.

## 2.5. CLIMATOLOGY AND URBAN PLANNING

The local differences in climate described in the previous sections, which are observable both within and directly above an urban conurbation, may also extend downwind in a plume-like fashion beyond its boundaries. Such urban–rural contrasts are invariably confined to the first few hundred metres of the atmospheric boundary layer and tend to fluctuate in both time and space. In these circumstances, the question inevitably arises as to the impact of these differences on the life of the inhabitants or, more importantly, the manner in which the change in climatic characteristics may either be avoided, mitigated, or even turned to advantage.

Landsberg (1981b) has observed that modern cities have tended to develop with little or no regard to the salient features of the prevailing climate. Indeed, urban planners have sometimes been guilty of perpetuating certain misconceptions, of which the most common appears to have been the location of industrial zones on the opposite side of the city to the prevailing, or most frequent, wind direction rather than that for stagnation-type weather conditions. Neiburger (1970) has also drawn attention to the frequent lack of consideration of adverse occurrences of light winds and temperature inversions. Topographic factors may be of particular importance in this respect, since cold drainage winds may allow air to settle in surface hollows, giving rise to low temperatures, fog formation and stagnation conditions.

Topographic considerations may also enter into the location of residential areas, for which sloping sites are to be favoured. In excessive sunshine the shaded slopes are superior, but in cold climates the sunlit slopes are to be preferred. When there are lakes and other large water bodies, advantage may be taken of lake breezes.

Within an urban area, the existence of the heat island lengthens the growing season, aids in fuel conservation and reduces the duration of snow cover. However, in summer the additional warmth not only causes extra stress for the inhabitants but also places additional load on air-conditioning plant. Parks and other open spaces may serve to reduce the strength of the heat island as well as providing a welcome amenity.

The layout of an urban area should itself reflect the local microclimate. For example, in desert regions, tall housing with small windows and narrow streets provides deep shade for most of the day. At coastal locations, streets may be arranged to enhance the inland penetration of the sea breeze. In contrast, in more moderate climates, both the orientation of the buildings and the width of the streets may be chosen to maximise both radiation and illumination. In addition, roof heights may be varied to increase turbulent ventilation.

Of the various aspects of urban climate described in Sections 2.2 to 2.4, perhaps the most important with regard to its impact upon the local community in general and infrastructure design in particular is the change in precipitation characteristics. One of the most thorough reviews of the consequences of such rainfall anomalies was carried out by Changnon *et al.* (1977) as part of the METROMEX experiment in St Louis, Missouri. Figure 2.4, which has been simplified from a diagram pertaining to St Louis presented by those authors, provides a summary of the more widely experienced impacts.

As noted in Section 2.3, the METROMEX experiment clearly identified local summer season increases in precipitation. In particular, the number, total depths and average rainfall rates of heavy storms were found to increase. Associated with this enhanced storm activity, more thunder and lightning were observed and more occurrences of hail were experienced. These changes were referred to by Changnon *et al.* (1977) as the direct impacts.

The more frequent occurrence of specific average rainfall rates within given durations was found to cause increases in both urban flooding and the surcharging of sewerage systems. As a result of the latter, the discharge of untreated effluents to local watercourses occur-

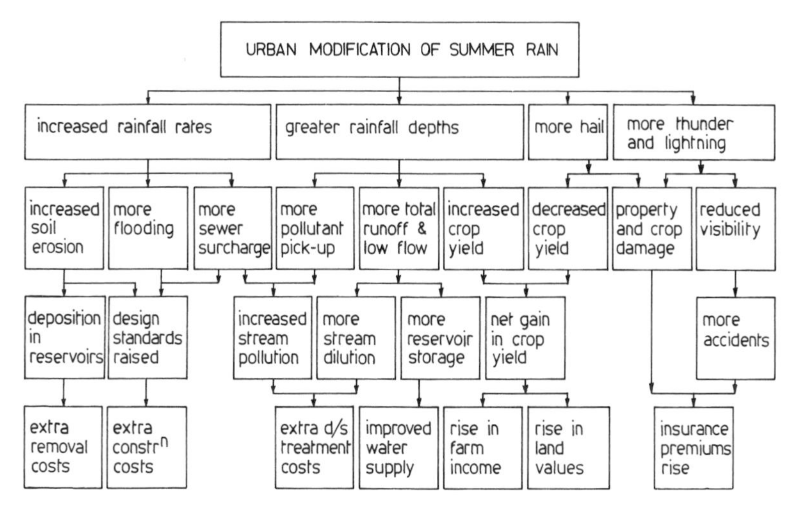


FIG. 2.4. Some consequences of rainfall anomalies in urban areas (modified from Changnon *et al.*, 1977, by permission of the Illinois State Water Survey).

red more frequently, thereby exacerbating stream pollution problems. Increasing the average rainfall rates for a given frequency of occurrence may also result in more soil erosion, which in turn may reduce the capacity of water supply and flood control reservoirs through sediment deposition. As a result of these secondary impacts, the community incurs extra costs both for the construction of new works to improved design standards and for the removal of sediment and the treatment of additional stream pollution.

A further series of secondary impacts may be associated with the increase in total rainfall depths. More scavenging of pollutants from the urban atmosphere takes place, which may ultimately contribute to pollutant loads in local streams. However, the greater rainfall depths result in larger runoff volumes, with summer low flows becoming notably larger. Such additional runoff may assist in the dilution of effluents, but probably not enough to avoid extra treatment costs, and may mitigate water supply problems by improving reservoir inflows at a time of peak demand. The contribution to soil moisture storage from the additional volume of rainfall can result in increased crop yields. In the St Louis area the latter effect was found to outweigh the reductions in crop yields caused by hail damage. The consequent net increase in crop production both improved farm income and caused a rise in the value of farmland affected by the precipitation anomaly. The increased incidence of hail, together with that of thunder and lightning, may increase the amount of damage to property as well as crops. The more frequent periods of reduced visibility caused by such weather conditions may have a profound effect upon transportation systems, resulting in a greater number of accidents and injuries, and possibly loss of life. The increase in insurance claims may ultimately cause a rise in premiums.

In summary, the precipitation anomaly identified by METROMEX was associated with an intensification of naturally occurring, moderate to heavy storms. In discussing its implications, Changnon *et al.* (1976) concluded that the urban-induced changes in rainfall were the result of the presence of the entire urban conurbation and not just one easily controllable human or industrial activity. According to Changnon *et al.* (1977), the net effect of the secondary and tertiary impacts of the anomaly is a considerable economic disbenefit. Unfortunately there appears to be little scope for avoiding the extra costs by more informed urban planning, although their magnitude might be reduced by judicious allocation of land use. With more cities assuming the proportions of Megalopolis, the changes in urban weather observable in the future are likely to exceed in magnitude those evaluated during the METROMEX experiment by a large margin.

## REFERENCES


ASHWORTH, J. R. (1929) On the influence of smoke and hot gases from factory chimneys on rainfall. *Quart. J. Roy. Met. Soc.*, **55,** 341–50.

ATKINSON, B. W. (1968) A preliminary examination of the possible effect of London's urban area on the distribution of thunder rainfall, 1951–1960. *Trans. Instn. Brit. Geogr.*, **44,** 97–118.

ATKINSON, B. W. (1969) A further examination of the urban maximum of thunder rainfall in London, 1951–1960. *Trans. Instn. Brit. Geogr.*, **45,** 97–120.

ATKINSON, B. W. (1970) The reality of the urban effect on precipitation: a case study approach. World Met. Organisation, Tech. Note No. 108, pp. 342–60.

BLENCH, B. J. R. (1963) Luke Howard and his contribution to meteorology. *Weather*, **18,** 83–92.

BRATER, E. F. (1968) Steps towards a better understanding of urban run-off processes. *Wat. Resour. Res.*, **4,** 335–48.

CHANDLER, T. J. (1965) *The climate of London* (Hutchinson, London) 292 pp.

CHANGNON, S. A. (1961) A climatological evaluation of precipitation patterns

over an urban area. Robert A. Taft San. Engng. Center, Tech. Rept. A62-5, pp. 37–67.
CHANGNON, S. A. (1968) The La Porte weather anomaly—fact or fiction. *Bull. Am. Met. Soc.*, **49,** 4–11.
CHANGNON, S. A. (1969) Recent studies of urban effects on precipitation in the United States. *Bull. Am. Met. Soc.*, **50,** 411–21.
CHANGNON, S. A. (1980) Is your weather being modified? *Proc. Am. Soc. Civ. Engrs., J. Irrig. Drain. Div.*, **106**(IR1), 37–48.
CHANGNON, S. A., SEMONIN, R. G. and HUFF, F. A. (1976) A hypothesis for urban rainfall anomalies. *J. Appl. Met.*, **15,** 544–60.
CHANGNON, S. A., HUFF, F. A., SCHICKEDANZ, P. T. and VOGEL, J. L. (1977) Summary of METROMEX, Vol. 1: Weather anomalies and impacts. Illinois State Wat. Survey, Bull. 62, 260 pp.
DETTWILLER, J. (1970) Incidence possible de l'activité industrielle sur les précipitations à Paris. World Met. Organisation, Tech. Note No. 108, pp. 361–2.
ELSOM, D. M. (1979) Air pollution episode in Greater Manchester. *Weather*, **34,** 277–86.
GEORGII, H. W. (1970) The effect of air pollution on urban climates. World Met. Organisation, Tech. Note No. 108, pp. 214–37.
HOLZMAN, B. G. and THOM, H. C. S. (1970) The La Porte precipitation anomaly. *Bull. Am. Met. Soc.*, **51,** 335–7.
HUFF, F. A. (1977) Urban effects on storm rainfall in Midwestern United States. Int. Assoc. Hydrol. Sci., Publ. No. 123, pp. 12–19.
JENKINS, I. (1969) Increases in averages of sunshine in central London. *Weather*, **24,** 52–4.
LANDSBERG, H. E. (1956) The climate of towns. In Thomas, W. L. (ed.), *Man's role in changing the face of the earth* (Univ. of Chicago Press) pp. 584–606.
LANDSBERG, H. E. (1961) City air—better or worse. Robert A. Taft San. Engng. Centre, Tech. Rept. A62–5, pp. 1–22.
LANDSBERG, H. E. (1981a) City climate. Ch. 3 of Landsberg, H. E. (ed.), *General climatology 3, World survey of climatology*, Vol. 3 (Elsevier, Amsterdam) pp. 299–334.
LANDSBERG, H. E. (1981b) *The urban climate* (Academic Press, New York) 275 pp.
LOWRY, W. P. (1977) Empirical estimation of urban effects on climate: a problem analysis. *J. Appl. Met.*, **16,** 129–35.
LUDWIG, F. L. (1970) Urban temperature fields. World Met. Organisation, Tech. Note No. 108, pp. 80–107.
MANLEY, G. (1958) On the frequency of snowfall in metropolitan England. *Quart. J. Roy. Met. Soc.*, **84,** 70–2.
MITCHELL, J. M. (1961) The thermal climate of cities. Robert A. Taft San. Engng. Centre, Tech. Rept. A62-5, pp. 131–45.
NEIBURGER, M. (1970) Air pollution considerations in city and regional planning. World Met. Organisation, Tech. Note No. 108, pp. 194–5.
OKE, T. R. (1974) Review of urban climatology, 1968–1973. World Met. Organisation, Tech. Note No. 134, 132 pp.

OKE, T. R. (1979) Review of urban climatology, 1973–1976. World Met. Organisation, Tech. Note No. 169, 100 pp.

OKE, T. R. (1982) The energetic basis of the urban heat island. *Quart. J. Roy. Met. Soc.*, **108,** 1–24.

OKE, T. R. and HANNELL, F. G. (1970) The form of the urban heat island in Hamilton, Canada. World Met. Organisation, Tech. Note No. 108, pp. 113–26.

PENWARDEN, A. D. (1973) Acceptable wind speeds in towns. *Building Sci.*, **8,** 259–67.

SHELLARD, H. C. (1959) The frequency of fog in the London area compared with that in rural areas of East Anglia and south-east England. *Met. Mag.*, **88,** 321–3.

UNWIN, D. J. (1980) The synoptic climatology of Birmingham's urban heat island, 1965–1974. *Weather*, **35,** 43–50.

# 3

# Rainfall Data for Flood Estimation and Drainage Design

## 3.1. INTRODUCTION

Since before the Industrial Revolution, the objective of stormwater drainage in towns and cities has been the rapid and efficient elimination of standing water from the built-up area. The attainment of this objective obviously depends primarily upon a knowledge of the characteristics of heavy rainfalls within short periods. The rate of accumulation of such knowledge has been limited largely by both the development of suitable instruments and the expansion of the rain-gauge network. As the design methods themselves have become more sophisticated, a need has arisen for more comprehensive analyses of the available data for the purpose of describing the temporal and spatial variations of storm events associated with given return periods.

During the nineteenth century, information on heavy falls of rain in short periods in the British Isles was collected by the British Rainfall Organisation, a group of volunteer observers whose data were collected and published by their founder, G. J. Symons, in an annual publication entitled *British Rainfall.* Even in the early days of this organisation, certain of its members were paying attention to drainage problems, and the following entry appeared in the observer's notes in the volume for 1862:

> Leek, Staffordshire.—We have an excellent system of public drainage in this town, and I have observed that the greater the number of days upon which the rainfall furnishes the sewers with a quantity of water equal to 30 gal/head on the population, the lower the number of deaths from zymotic disease,—this rule held good for 3 years. . . . At the end of any quarter I am able by the Rain Gauge to

> inform the Registrar, with tolerable accuracy, the number of deaths he has registered from diseases of the above mentioned class.

The British Rainfall Organisation published their first table of heavy rainfalls in short periods in 1888, covering the decade 1879–1888. These data, which were classified as either 'noteworthy' or 'exceptional', may be regarded as one of the first attempts to compile the basis for a rainfall depth–duration–frequency (DDF) relationship. A further review in 1908, covering the period 1879–1908, distinguished between three classifications: 'noteworthy', 'remarkable' and 'very rare'. Since these data were obtained with ordinary storage gauges, the depths recorded within the shorter periods must have been seriously underestimated. With the introduction of autographic rainfall recorders during the 1920s, more reliable data began to be acquired and more formal statistical analysis permitted a more precise definition of relative frequency. The variety of analytical expressions that have been proposed to describe both rainfall intensity–duration–frequency (IDF) and DDF relationships are considered in greater detail in Section 3.2.

The relationships discussed in Section 3.2 are statistical abstractions of point rainfall, i.e. precipitation as observed at a single raingauge. These data are sufficient to allow the following generalisations to be made about the characteristics of storm rainfall:

(1) as storm duration increases, the average rainfall intensity decreases for any given frequency of occurrence; and
(2) as the frequency of occurrence decreases, the average rainfall intensity increases for any given duration.

These generalisations relate to temporal variations only. In order to evaluate the spatial characteristics of storm rainfall, data from networks of raingauges must be studied. From the results of such investigations, a third feature of heavy rainfall has become well established by observation:

(3) the greater the area covered by a storm, the lower the average rainfall intensity compared with the maximum point rainfall intensity recorded within the storm boundaries.

This reduction in the areal average rainfall intensity with respect to the maximum intensity observed may be described for individual events by means of a rainfall depth–area–duration (DAD) relationship. However, when attempting to adjust a design (point) depth to allow for spatial distribution, the ratio of the areal average to the point average rainfall intensities corresponding to the same frequency of occurrence is required. This ratio, which is referred to as the areal reduction factor (ARF), is often confused with the rainfall DAD relationship, although both are useful for distinctly different purposes, as described in Section 3.3. In addition, since many design flood estimating methods require a single, representative rainfall input for the whole catchment area, techniques for averaging the records from several raingauges are often required. A review of the several available approaches is also presented in Section 3.3. Whereas the average rainfall intensity within a given duration corresponding to a predetermined frequency of occurrence is sufficient to construct a design storm for the simpler flood estimation methods giving a peak rate of flow only, the temporal variation of rainfall is required to synthesise a discharge hydrograph. Initially, such storm profiles were envelope curves constructed by integrating the rainfall DDF relationship over time. However, with the accumulation of data from autographic raingauges, analysis of selected storm events became feasible. The results obtained are summarised in Section 3.4.

As the sophistication of mathematical models of the relationship between rainfall and runoff has increased, greater demands have been placed on the meteorologist to provide more comprehensive analytical descriptions of storm rainfall. If the mathematical model is capable of accepting what is known as a distributed input, i.e. different zones of the catchment area being modelled having different storm profiles as opposed to a single 'lumped' profile, the question arises as to whether the speed and direction of movement of any design storm should be taken into account. This topic is considered further in Section 3.5.

## 3.2. RAINFALL DEPTH–DURATION–FREQUENCY RELATIONSHIPS

Although the meteorologist prefers to work with the total depths of rainfall observed within different durations, the engineer has traditionally adopted average intensity, i.e. the quotient of total depth and duration, as his primary rainfall variable. The rainfall IDF relationship is therefore a transformation of the rainfall DDF relationship, with the implicit understanding that the intensity of rainfall is constant during the specified duration.

The evolution of rainfall DDF relationships may be conveniently illustrated by considering the work which has been carried out in the British Isles, where Lloyd-Davies (1906) was among the first to carry out an analysis of rainfall records for the purposes of stormwater drainage design. Using records from Edgbaston Observatory, Birmingham, for the years 1900–1904, Lloyd-Davies compiled a table showing the total numbers of storms with given average rainfall intensities for a range of durations. These data were represented by a relationship later referred to as the Birmingham curve, which was intended to include all but the abnormal storms, provision for which would be outside economic limits:

$$i = a/(t + b) \tag{3.1}$$

where $i$ is the average rainfall over duration $t$, and $a$, $b$ are constants. With the intensity measured in in/h and the duration in min, $a = 40$ and $b = 20$.

Similar analyses to those carried out by Lloyd-Davies were subsequently undertaken by municipal engineers working for several other British local authorities. Curves of the form of eqn (3.1) were fitted to the available local rainfall records, but inevitably produced different values of the constants $a$, $b$. For example, using the same units of in/h and min for $i$ and $t$, Roseveare (1930) reported $a = 20$ and $b = 8$. The variety of such relationships in use, coupled with the lack of consistency in the frequency of occurrence attached to different curves, created difficulties for central government when details of surface water drainage schemes from different parts of the country were submitted for grant aid. This problem was considered by the Ministry of Health Departmental Committee on Rainfall and Runoff (1930), who recommended the adoption of a standard working curve of the form of eqn (3.1) in which $a = 30$ and $b = 10$ for $5 \leqslant t \leqslant 20$ min, and the constants for the Birmingham curve were substituted for $20 \leqslant t \leqslant 100$ min.

These relationships, which became known as the Ministry of Health formulae, were based upon records for 14 stations for the period 1921–1927 relating to the number and intensity of storms of 10, 15, 30 and 60 min duration, and 7 years of data from an additional 7 stations for the 4 heaviest storms of 5, 10 and 15 min duration. The formulae were stated to cover all except the two or three heaviest storms during the 7-year period for the average of all the stations. A more precise definition of frequency of occurrence had to await the publication of the results of a more extensive study by Norris (1948). Using autographic rainfall records from 40 stations for up to 21 years of data (1925–1946), Norris was able to show that the Ministry of Health formulae corresponded to a frequency of once a year.

An alternative form of rainfall intensity–duration relationship to eqn (3.1) was proposed by Escritt (1950):

$$i = a/t^m \tag{3.2}$$

where $a$ is a constant and $m$ an exponent. By drawing a mean curve through the results obtained by previous investigators, notably Bilham (1935) and MacLean (1945), an alternative working curve to the Ministry of Health formulae was produced using $a = 5{\cdot}9$ and $m = 0{\cdot}625$. These values provide a tolerable agreement with those obtained from a comprehensive analysis of 35-year autographic rainfall records for Cork by Dillon (1954). Using the more general form of relationship

$$i = a(f/t^3)^m \tag{3.3}$$

where $f$ is the frequency of occurrence, Dillon quoted values of $a = 6{\cdot}0$ and $m = 0{\cdot}2$, making the power of the duration term 0·6. Equation (3.3) was said to apply to any duration between 5 min and 24 h.

Clearly, a more general form of rainfall IDF relationship may be obtained by combining eqns (3.1) and (3.3) to give

$$i = af^m/(t + b)^n \tag{3.4}$$

where $n$ is another exponent. This expression has been applied to Australian rainfall data by McIllwraith (1953) and to records from the British Isles by Collinge (1961). The latter author was able to show that the coefficients $a$, $b$ were a function of the average annual rainfall and thunderstorm frequency, expressed as the number of days per year on which thunder was heard. An indirect method was proposed for estimating the exponent $n$, and $m$ was found to be a function of $a$ and the logarithm of the duration. The method was recommended for use in estimating average rainfall intensities for $10\ \text{min} \leqslant t \leqslant 5\ \text{h}$ and any frequency of occurrence, but attracted little support, partly because of the scarcity of records of thunderstorm frequency, and partly owing to the complexity of the procedure required to obtain values of the constants and exponents. The simpler forms of expression, typified by the Ministry of Health formulae, continued to dominate in engineering practice along with eqn (3.2) and another form of rainfall DDF relationship suggested by MacLean (1945):

$$R = a - b \log f \qquad (3.5)$$

where $R$ is the rainfall depth, $f$ the frequency in number of events per year and the constants $a$, $b$ are functions of duration. Equation (3.5) appears to have been intended more for overseas use.

The above discussion has concentrated primarily on the efforts of the engineers to derive working curves for average rainfall intensities that could be applied to drainage design problems. In parallel with these developments, meteorologists continued to build upon the foundations laid by the British Rainfall Organisation in both recording and classifying heavy falls of rain within short periods. The British Rainfall Organisation was transferred to the Meteorological Office in 1919. Following the introduction of autographic rainfall recorders during the 1920s, a more precise definition of the categories 'noteworthy', 'remarkable' and 'very rare' became possible. The first decade of data from such gauges, relating to the period 1925–1934 at 12 sites, was analysed by Bilham (1935), who used a rainfall DDF relationship of the form

$$N = 1{\cdot}25(t/60)(R + 0{\cdot}1)^{-3{\cdot}55} \qquad (3.6)$$

where $N$ is the number of days in 10 years during which a rainfall of $R$ (in) fell within a duration of $t$ (min) or less. Equation (3.6) was intended to be applicable for durations between 5 and 120 min. As a result of Bilham's analysis, the following definitions were adopted for rainfall classification:

| *Classification* | *N* | *Return period (years)** |
|---|---|---|
| Noteworthy | 1·0 | 10 |
| Remarkable | 0·25 | 40 |
| Very rare | 0·0625 | 160 |

* Return period is the reciprocal of the frequency (no. of events/year).

During the 1960s a comprehensive reappraisal of Bilham's work was carried out by the Meteorological Office. As reported by Holland (1964), eqn (3.6) was found to have withstood the test of time remarkably well. Apart from increasing the constant from 1·25 to 1·39 and straightening the curve when the duration was less than 1 min, the major recommendation involved replacing eqn (3.6) by the following expression when the average rainfall intensity exceeded 1·25 in/h:

$$N = R \exp(1 - 0{\cdot}8R/t)(R + 0{\cdot}1)^{-3{\cdot}55} \qquad (3.7)$$

Perhaps the major criticism that could be levelled at the Bilham formula in both its original and modified forms is its failure to take account of regional variability in the frequency of heavy rainfall. In presenting eqn (3.6), Bilham (1935) noted that areas with a high annual rainfall are more liable to prolonged heavy rains than areas with a low annual average rainfall. Evidence of regional departures from the Bilham formula was provided by Ashworth and O'Flaherty (1974) and is further illustrated by Fig. 3.1, which shows contours of percentage departure from eqn (3.6) for a typical duration and return period. This figure shows that, although much of England is well represented by the Bilham formula, substantial local anomalies exist, which could only be taken into account by the inclusion of representative data within the framework of a more thorough regional analysis.

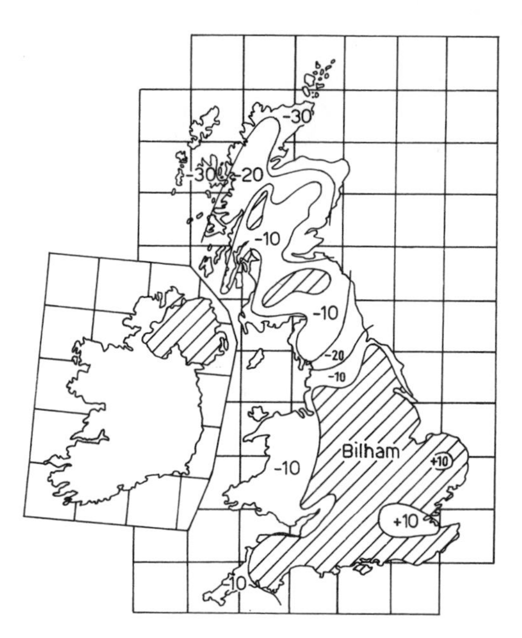


FIG. 3.1. Simplified contours of the 15 min, 1-year rainfall depth expressed as a percentage of the value computed from the Bilham formula.

Such an investigation was subsequently undertaken by the Meteorological Office as part of the work connected with the preparation of the United Kingdom Flood Studies Report (Natural Environment Research Council, 1975).

The method recommended in the Flood Studies Report for the construction of the rainfall DDF relationship for any given location in the United Kingdom was based upon the analysis of a substantial data bank which included:

(i) 600 daily raingauges having an average of 60 years of record;
(ii) a further 6000 daily raingauges operating during 1961–1970;
(iii) short-duration rainfall data from 100 sites;
(iv) records from dense networks of autographic raingauges at two sites; and
(v) data on rainfall within very short durations obtained from a small number of Jardi rainfall recorders.

The approach depends upon two reference rainfall depths: the 2-day and the 60 min totals corresponding to a 5-year return period. Initially, the point of interest is located by means of its national grid reference. Maps are then employed to determine the 2-day, 5-year rainfall depth, which for convenience is denoted by the abbreviation M5–2 day, and the ratio, $r$, of the M5–60 min to the M5–2 day totals. Tables are provided which permit the M5–$t$ duration rainfall depth to be interpolated as a percentage of the M5–2 day value, given the ratio $r$, for durations up to 48 h. A further two tables, one for England and Wales and the other for Scotland and Northern Ireland, give the factors, M$T$/M5, to convert the M5–$t$ duration depth to the M$T$–$t$ duration total, where $T$ is the required return period.

With the launching of the Wallingford Procedure for the design and analysis of urban storm drainage systems (National Water Council and Department of the Environment, 1981), a computerised version of the Flood Studies Report method for constructing rainfall DDF curves was introduced. In this approach, the M5–2 day map was replaced by another showing M5–60 min rainfall depths, since the latter are more pertinent to urban drainage problems.

The above-mentioned methods for estimating the $T$-year, $t$-duration rainfall depth apply specifically to the United Kingdom. For other parts of the world similar relationships can be constructed from an analysis of the records obtained from autographic rainfall recorders. Unfortunately, especially in the more arid regions, such records (if available at all) are generally for very short periods of observation. Where a design rainfall depth is required corresponding to a return period which equals or exceeds the record length, more reliable estimates are probably obtained from generalised relationships based upon the large numbers of long rainfall records from the United States of America. The latter cover a sufficiently wide range of climates for analogues to be selected for other parts of the world. The use of data from Arizona and California to derive design rainfall depths for airport drainage in North Africa by the US Weather Bureau (1954) provides a ready example of this approach.

The development of generalised rainfall DDF relationships based upon American data initiated by Hershfield and Wilson (1957) and Hershfield *et al.* (1955) was continued by Reich (1963) and Bell (1969). The success of this approach depends primarily upon the assumption that extreme rainfall events having a duration of (say) 2 h or less are usually obtained from similar structures of storm systems such as convective cells which are known to have comparable physical properties in many parts of the world. The generalised relationships are often presented in graphical form, but Bell (1969), using data from 157 raingauges in the United States with more than 40 years of record, has shown that the $T$-year, $t$-minute depth, $R^t_T$, may be estimated using the expression

$$R^t_T = R^{60}_2(0{\cdot}35 \ln T + 0{\cdot}76)(0{\cdot}54t^{0{\cdot}25} - 0{\cdot}5) \qquad (3.8)$$

Equation (3.8), which applies to return periods between 2 and 100 years and durations between 5 and 120 min, depends upon depth–duration ratios being independent of frequency and depth–frequency ratios being independent of duration. The reference rainfall depth, which in eqn (3.8) is the 2-year, 60 min depth, may either be estimated from local rainfall records or computed from the equations

$$\begin{aligned} R^{60}_2 &= 0{\cdot}17\bar{R}M^{0{\cdot}33};\ 0 < M \leqslant 50,\ 1 < N \leqslant 80 \\ &= 0{\cdot}617\bar{R}^{0{\cdot}67}M^{0{\cdot}33};\ 50 < M \leqslant 115,\ 1 < N \leqslant 80 \end{aligned} \qquad (3.9)$$

where $\bar{R}$ is the mean annual maximum observer-day rainfall total (mm) and $M$ is the mean annual number of thunderstorm-days. Equation (3.9) is again based upon long-term records from the United States.

More recently, the following alternative expression to eqn (3.8) has been suggested by Hargreaves (1981):

$$R^t_T = KT^{1/6}t^{1/4} \qquad (3.10)$$

where $K$ is a constant which must be derived for each location. Equation (3.10) has been found to apply to $30\,\text{min} \leqslant t \leqslant 4$ days and $5 \leqslant T \leqslant 100$ years.

## 3.3. SPATIAL VARIATIONS IN RAINFALL DEPTHS

Since the rise and fall in streamflow at a particular river cross-section represents an integration of both temporal and spatial variations in storm precipitation, the drainage engineer is more concerned with the average rainfall over an area than with totals at individual gauges. Indeed, the majority of rainfall–runoff models in current use require as input data a single representative time sequence of rainfall intensities. Where the catchment area is sufficiently large to have three or more raingauges located within or adjacent to its watershed, the problem arises of compiling a suitable rainfall hyetograph from the available information. There are three methods of determining the average depth of rainfall over an area that have been in widespread use for many years: the arithmetic mean, the Thiessen method and the isohyetal method.

Of these three approaches, the arithmetic mean, which merely involves averaging the observed rainfall totals from each gauge within the catchment area, is undoubtedly the simplest. Nevertheless, the method is adequate in relatively flat topography where the raingauges are uniformly distributed and individual observations do not depart widely from the mean. Where the distribution of gauges is non-uniform, a series of weighting factors can be obtained by applying the Thiessen method. As introduced by Thiessen (1911), this technique is based on a geometrical construction in which:

(a) the raingauge sites are plotted on a map and connection lines are drawn;
(b) the perpendicular bisectors of the lines connecting each pair of adjacent gauges are constructed to form polygons around each site;
(c) weighting factors are computed for each gauge equal to the proportion of the total catchment area enclosed within the polygon and the watershed; and
(d) the average rainfall is computed from the sum of the products of gauge weight and observed rainfall depth.

In effect, the Thiessen method assumes a linear variation in rainfall totals between gauges, and assigns segments of the catchment area to the nearest station in the network. The technique is somewhat inflexible in that any change in the number of gauges requires the computation of a new set of weights. Determination of the latter can be accomplished more efficiently by the use of computer-based procedures, such as those described by Diskin (1969, 1970). An approach similar to the Thiessen method which avoids the measurement of areas has been proposed by Pande and Al-Mashidani (1978). Based upon a construction of radial lines, the latter method requires at least three gauges.

A further important limitation of the Thiessen method is the absence of opportunity for the analyst to incorporate his knowledge of storm morphology and orographic influences. The isohyetal method, in which contours of equal rainfall depth, referred to as isohyets, are interpolated on a map showing the observed totals at each location, provides this extra flexibility and is generally regarded as the most accurate approach. Nevertheless, the isohyetal method is both laborious and time-consuming, particularly where large numbers of storm events are to be processed, and several authors have investigated the possibility of applying objective computer-based procedures for constructing isohyetal maps. The latter involve the use of surface-fitting techniques in which the rainfall is defined mathematically in terms of $x$ and $y$ coordinates. Given the observed rainfall depths, such techniques involve the derivation of a set of basic equations whose coefficients must be computed from the data. Given these coefficients, the rainfall total at any point of interest can be calculated and average depths determined by either analytical or numerical integration. Particular attention has been paid to two techniques that are well adapted to the irregularly spaced data points found in raingauge networks: trend surface analysis and multiquadric analysis.

The application of trend surface analysis to the description of isohyetal patterns has been discussed by Amorocho and Brandstetter (1967), Unwin (1969), Chidley and Keys (1970), Mandeville and Rodda (1970) and Singh (1976) among others. The surface may be represented by the basic equation

$$R = \sum^{k} a_i f_i \tag{3.11}$$

where $a_i$ are coefficients, the $f_i$ functions of latitude, longitude and altitude, and $R$ are the rainfall observations. For example, the $f_i$ may be expressed in terms of the $x$ and $y$ coordinates to give an expression of the form

$$R = a_0 + \sum^{k} a_i x^i + \sum^{k} b_i y^i + \sum^{k} c_i x^i y^{k-i} \tag{3.12}$$

where the coefficients $a_0$, $a_i$, $b_i$, $c_i$, $i = 1, 2, \ldots, k$, are determined by the method of least squares. The more complex the trend function, the more representative will be the surface, but the greater the possibility that numerical problems will be encountered in solving for the coefficients. Indeed, Singh (1976) has concluded that for computing mean areal rainfalls for modelling purposes, trend functions offer no advantages over the well-established methods.

In contrast to trend surface analysis, which provides an equation giving a best least-squares fit to the rainfall observations, multiquadric analysis produces a surface which fits all data points exactly. With the latter technique, the surface is represented by a summation of individual quadric surfaces, such as circular hyperboloids and circular paraboloids. For example, Shaw and Lynn (1972) discussed the use of a special case of the former in which the surface is a right circular cone, which for $k$ raingauges gives rise to the following system of $k$ equations:

$$R_i = \sum_{j=1}^{k} C_j \sqrt{[(x_j - x_i)^2 + (y_j - y_i)^2]} \tag{3.13}$$

where $R_i$ is the rainfall depth at the $i$th site whose coordinates are $x_i$, $y_i$ and the $C_j$ are coefficients to be determined. In a follow-up study, Lee *et al.* (1974) showed that computational difficulties occurred when the more complex multiquadric hyperboloids were used, and that multiple cones, as given by eqn (3.13), were to be preferred for their simplicity and efficiency.

Although isohyetal maps provide a visual impression of the magnitude of a storm, the intercomparison of events is more conveniently carried out by the use of rainfall depth–area–duration (DAD) curves. In effect, such curves show the relationship between the average rainfall depth within a given isohyet and the area enclosed by that isohyet for a constant duration. A variety of empirical formulae has been proposed to describe such relationships, ranging from fractional powers of the area to logarithmic and exponential, a comprehensive listing of which has been presented by Court (1961). Alternatively, the data may be expressed in the form of a series of reduction factors (see Bell, 1976):

$$\mathrm{RF} = R_1/R_2 \tag{3.14}$$

where $R_1$ is the maximum areal rainfall within a storm event over a given area and duration, and $R_2$ is the maximum point rainfall within the same storm for the same duration. The ratio given by eqn (3.14) is often referred to as a 'storm-centred' reduction factor, since $R_1$ is determined by the isohyetal pattern, which is inevitably centred on the maximum, $R_2$. Consequently, the area covered by the analysis varies according to the morphology of the storm. This type of information is therefore unsuitable for use in drainage design, where the engineer is concerned with a specific catchment area and the variations in rainfall that are likely to occur within its boundaries. For the latter purpose, a 'fixed-area' reduction factor may be defined by the ratio

$$\mathrm{ARF} = R_3/R_4 \tag{3.15}$$

where $R_3$ is the mean areal rainfall for a given duration and return period and $R_4$ is the mean point rainfall for the same duration and return period in the same area. This ratio should more properly be referred to as the areal reduction factor.

The important distinction that must be drawn between RF and ARF is the equality in return periods implicit in the latter. In any given storm, $R_2$ will invariably have a larger return period than $R_1$, so that the fixed-area reduction factors are usually larger than the storm-centred ratios. Attempts to relate ARF values to characteristics of particular storms are therefore erroneous.

Among the first comprehensive evaluations of ARF values was that of Holland (1967), who used data from a network of autographic raingauges set out at approximately 1 km intervals over an area near Cardington in England. The results obtained were summarised in the form of the equation

$$\mathrm{ARF} = 1 - [3\sqrt{A}/\Gamma^{-1}(t)] \tag{3.16}$$

where $A$ is the area (ha), $t$ is the duration (min) and $\Gamma^{-1}$ is the inverse gamma function. Equation (3.16) was intended to apply to storm durations between 2 min and 2 h and was subsequently presented in graphical form by the Meteorological Office (1968).

The work of Holland (1967) has now been superseded by that presented in the United Kingdom Flood Studies Report (Natural

Environment Research Council, 1975). These revised ARF values were computed according to a modified definition:

$$\mathrm{ARF}' = \overline{(R_5/R_6)} \tag{3.17}$$

where $R_5$ is the rainfall at any point within a given area occurring at the time of the annual maximum areal rainfall, and $R_6$ is the annual maximum rainfall for the same point, the same duration and the same year (but not necessarily the same time) as $R_5$. The bar denotes an averaging procedure over both points contained within the area and a number of years of record. The results obtained, which apply to a wider range of durations and areas than those computed by Holland (1967), are presented in Fig. 3.2.

Since the Flood Studies Report definition of ARF is based upon mean annual maxima, any systematic variation of such factors with return period cannot be identified. However, Bell (1976), using the definition given by eqn (3.15), found a statistically significant trend for ARF values to decrease with increasing return period. The factors presented in Fig. 3.2 therefore tend to give conservatively high areal rainfall totals for return periods in the range 10–100 years, amounting

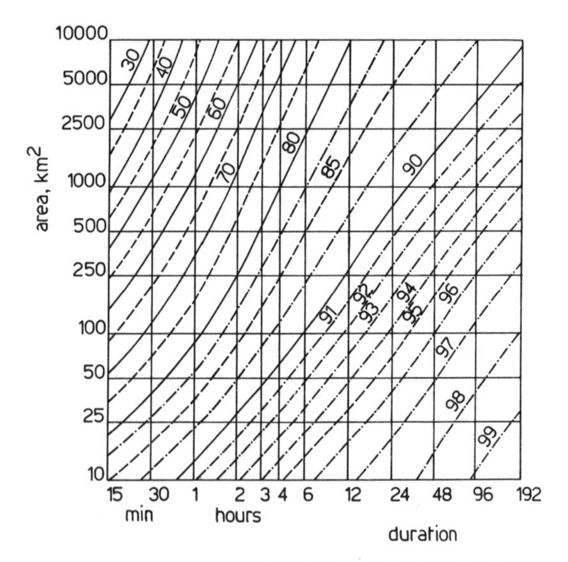


FIG. 3.2. Flood Studies Report areal reduction factors (redrawn from NERC, 1975, by permission).

to 10% at durations of 1–2 h and 5% at 24 h. However, these variations are small compared with the sampling inaccuracies in the derivation of ARF values. Bell (1976) also noted that, based upon evidence from both the United Kingdom and the United States, fixed-area ARF values appear to vary little with geographical location, so that Fig. 3.2 may have a wider application, at least as a first approximation.

## 3.4. TEMPORAL VARIATIONS IN RAINFALL DEPTHS

Since changes in streamflow are a reflection of temporal as well as spatial variations in rainfall intensity, the drainage engineer also requires information on storm profiles. Three distinct approaches have been employed in order to construct such profiles, namely:

(1) by integration of a rainfall DDF relationship;
(2) by the extraction of representative time variations from storms recorded by autographic raingauges; and
(3) by the approximation of the profile by a simple geometrical figure.

Ross (1921) appears to have been among the first to apply method (1) above. The procedure advocated by Ross for constructing a profile from a rainfall DDF curve is summarised in Table 3.1 using the Ministry of Health formulae for purposes of illustration. Assuming that the profile is to be defined at 5 min time increments, columns 1–3 of Table 3.1 give both the average rainfall intensities and total depths for durations between 5 and 50 min. Column 5 then gives the rainfall totals within successive increments, which are obtained by subtracting the 5 min total in column 3 from the 10 min depth, the 10 min total from the 15 min depth, and so on. The totals for successive increments given in column 5 are converted into average intensities in column 6 to give the required rainfall profile.

The most obvious and dubious property of any storm profile produced using this procedure is the occurrence of the peak ordinate during the first increment. Although not impossible, such a profile must be regarded as atypical, certainly for temperate climates. Moreover, with the exception of the small rise in intensity at the fifth time increment (which can be attributed to the change in the constants of the equation at a duration of 20 min), rainfall rates continue to

**Table 3.1. Construction of a storm profile from a rainfall depth–duration–frequency relationship: an illustration using the Ministry of Health formulae**

| *Duration (min)* | *Intensity (mm/h)* | *Depth (mm)* | *Increment number* | *Depth (mm)* | *Intensity (mm/h)* |
|---|---|---|---|---|---|
| 5 | 50·8 | 4·23 | 1 | 4·23 | 50·8 |
| 10 | 38·1 | 6·35 | 2 | 2·12 | 25·4 |
| 15 | 30·5 | 7·62 | 3 | 1·27 | 15·2 |
| 20 | 25·4 | 8·47 | 4 | 0·85 | 10·2 |
| 25 | 22·6 | 9·41 | 5 | 0·94 | 11·3 |
| 30 | 20·3 | 10·16 | 6 | 0·75 | 9·0 |
| 35 | 18·5 | 10·78 | 7 | 0·62 | 7·4 |
| 40 | 16·9 | 11·29 | 8 | 0·51 | 6·1 |
| 45 | 15·6 | 11·72 | 9 | 0·43 | 5·2 |
| 50 | 14·5 | 12·10 | 10 | 0·38 | 4·5 |

Note: For durations $5 \leqslant t \leqslant 20$ min, $i = 762/(t+10)$; for $t > 20$ min, $i = 1016/(t+20)$, with intensity $i$ in mm/h.

decrease as the storm proceeds. Doubts concerning the authenticity of such a profile led several authors, including Hawken (1921) and Judson (1933), to suggest an arbitrary rearrangement in the rainfall intensities for successive time increments. Ormsby (1933) presented two such profiles based upon the Ministry of Health formulae, one of which reached its peak after one-third of its duration and the other after one-half. The synthetic profile derived by Keifer and Chu (1957) for the Chicago area, which was based upon a different form of rainfall DDF curve, was constructed to peak at three-eighths of the duration.

Despite the widespread and continued use of this approach for constructing storm profiles (see, for example, Chien and Sarikelle, 1976), the method should be discarded because such profiles inevitably have a lower frequency than the rainfall DDF curve from which they were derived. As demonstrated by Frederick (1978) among others, the joint occurrence of large rainfall totals within different durations of the same storm has a low probability.

With the refinement of methods for stormwater drainage design during the early 1960s, a need arose for a range of more representative storm profiles. The results from the Cardington rainfall experiment (Holland, 1967) were used to construct a series of design profiles for use with what is now known as the Transport and Road Research Laboratory Hydrograph Method (see Road Research Laboratory,

1963). These profiles were derived by expressing their ordinates as a proportion of the peak and then averaging the dimensionless ordinates obtained from separate events. The resultant profile, which had a slight asymmetry, was then scaled in proportion to the 15 min average rainfall intensities computed from the Bilham formula in order to produce storm patterns corresponding to different return periods. Subsequently, King (1967) fitted two equations to these profiles:

$$i = K \exp(-t^{0.75}/4) \qquad (3.18)$$

$$i = K \exp(-t/7) \qquad (3.19)$$

where $i$ is the instantaneous rainfall rate at time $t$ min measured from the peak ordinate, and $K$ is a constant depending upon return period.

The rainfall profiles quoted by the Road Research Laboratory (1963), a tabulation of which was later provided by Watkins (1966), continued in use until the publication of the Flood Studies Report (Natural Environment Research Council, 1975). This revision was based upon the analysis of a wider range of storm events having durations up to four rain-days. Each storm was centred on the shortest duration giving at least half the rainfall total, resulting in a set of mean profiles that were symmetrical but varied in amplitude. These profiles were then ranked according to 'percentile peakedness', i.e. the percentage of occasions when storms were less peaked than a given mean profile. Separate analyses were carried out for summer (May to October) and winter (November to April) storms, and in both cases variations in profile with storm duration and the return period of the average intensity of rainfall during the storm were found to be relatively insignificant.

Typical examples of Flood Studies Report storm profiles of different percentiles of peakedness are presented in Figs 3.3 and 3.4. In each case, these profiles are compared with the new design storms recommended for use with the Transport and Road Research Laboratory Hydrograph Method as given in the 2nd edition of Road Note 35 (see Transport and Road Research Laboratory, 1976). The latter represent a simplified version of the Flood Studies Report profiles in which variations with duration and peakedness have been ignored, as a result of which marked variations in profile shape are possible, as shown by Figs 3.3 and 3.4. These simplified profiles are therefore recommended only for preliminary calculations, and should not be used where storage effects are important, such as in the design of balancing tanks.

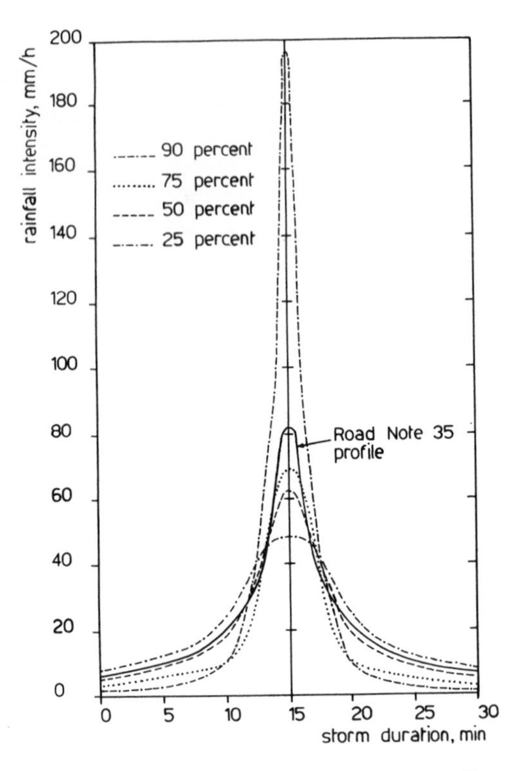


FIG. 3.3. Flood Studies Report 30 min, 1-year storm profiles compared with the simplified Road Note 35 profile.

Further discussion of the differences between the profiles has been presented by Folland (1978).

The storm profiles resulting from the Cardington experiment and the analyses carried out for the Flood Studies Report as described above were derived from complete storm events. In contrast, Pilgrim *et al.* (1969) and Pilgrim and Cordery (1975) have preferred to concentrate their attention on intense bursts of rainfall within longer storms in order to derive design storm profiles for Sydney, Australia. Those authors presented a ranking procedure for constructing temporal rainfall patterns having an average variation in intensities.

The differences between these two approaches for abstracting typical storm profiles from autographic raingauge records give rise to specula-

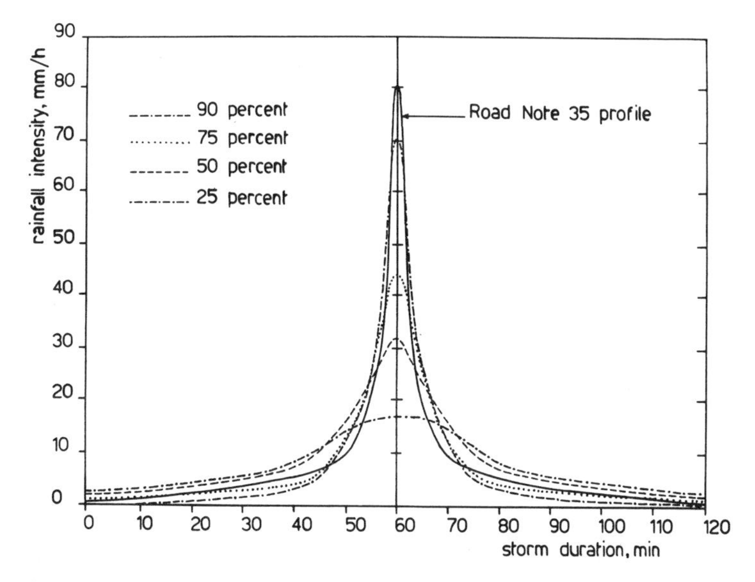


FIG. 3.4. Flood Studies Report 120 min, 1-year storm profiles compared with the simplified Road Note 35 profile.

tion as to the consequences of using one or the other to produce inputs for a rainfall–runoff model. However, the storm profile is only one of the variables that has to be chosen by the designer; other parameters, such as the antecedent wetness condition of the catchment, also have an important bearing on the relationship between the return period of the peak runoff rate and the return of the storm rainfall depth. This aspect of flood estimation methods is discussed more fully in Chapter 4.

The complexities of abstracting representative temporal patterns from rainfall records has led some authors to search for simpler shapes of design hyetographs based upon geometrical figures, such as the triangular storm profile suggested by Yen and Chow (1980). Those authors showed that, once the duration and average intensity of the storm had been specified, only the first moment of the area of the hyetograph was required to determine its shape. However, the success of this approximation again depends upon the flood estimation method in which the profile is applied and the manner in which the other significant design variables are chosen.

## 3.5. STORM MOVEMENT

Any design storm that is employed as the input to a flood estimation method is invariably assumed to remain stationary with respect to the catchment area under study. Nevertheless, laboratory studies in which artificial rainfall has been applied to elementary catchments consisting of sloping planes have demonstrated that upstream movement parallel to the main channel reduces, but downstream movement enhances, the peak runoff rate for a stationary storm (see, for example, Yen and Chow, 1969). Unfortunately, as noted by Shearman (1977), there is a dearth of statistical information on the speed and direction of storm movement and the areal extent of storms. However, studies by Felgate and Read (1975), Shearman (1977) and Marshall (1980) have provided some preliminary, if conflicting, results using statistical techniques involving the cross-correlation of the records from pairs of raingauges in a network.

Shearman (1977), who studied the data from 15 autographic raingauges in the Surrey and Greater London area of southern England, was unable to find support for the empirical rule in meteorology that storms move at the wind speed observed at the 700 mbar level in the atmosphere. In contrast, Marshall (1980), who analysed records from the Cardington experiment and a similar network at Winchcombe, found that storms moved in the same general direction as the 700 mbar wind speed, only more slowly. Both authors found that storm speeds were well in excess of design flow velocities in conventional sewerage systems. Shearman reported that 60% of storms moved faster than 15 m/s, and Marshall quoted mean storm velocities for both Cardington and Winchcombe of about 11·5 m/s.

The results obtained by Shearman have been used in numerical experiments by Sargent (1981, 1982) to investigate the importance of storm movement in the design of sewer networks. In this study, synthetic storms of 2 h duration having circular isohyets and a typical temporal pattern were moved at different speeds and in different directions across a series of rectangular catchment areas having notional drainage systems with different design flow velocities. The results obtained showed that, although peak runoff rates could be increased by downstream storm movement, the amount of enhancement was negligible, being a maximum of about 1% when storm velocity was equal to flow velocity. In contrast, when the apparent storm duration was less than the time of flow, substantial reductions in peak runoff rates compared with those of stationary storms could be produced with catchments between 2 and 10 km$^2$ in area. This tendency towards overdesign caused by storm movement is obviously worthy of further investigation, perhaps using the more sophisticated models of storm sequences such as those described by Sieker (1977) for the Hamburg area and Amorocho and Wu (1977) for cyclonic rainfall in northern California.

## REFERENCES

AMOROCHO, J. and BRANDSTETTER, A. (1967) The representation of storm precipitation fields near ground level. *J. Geophys. Res.*, **72,** 1145–63.

AMOROCHO, J. and WU, B. (1977) Mathematical models for the simulation of cyclonic storm sequences and precipitation fields. *J. Hydrol.*, **32,** 329–45.

ASHWORTH, R. G. and O'FLAHERTY, C. A. (1974) Validity of the modified Bilham equation. *J. Instn. Munic. Engrs.*, **101,** 43–50.

BELL, F. C. (1969) Generalised rainfall–duration–frequency relationships. *Proc. Am. Soc. Civ. Engrs., J. Hydraul. Div.*, **95**(HYl), 311–27.

BELL, F. C. (1976) The areal reduction factor in rainfall frequency estimation. Institute of Hydrology, Wallingford, Rept. No. 35, 58 pp.

BILHAM, E. G. (1935) Classification of heavy falls in short periods. Revision of the curve showing the lower limits of 'noteworthy', 'remarkable' and 'very rare' falls. In *British Rainfall*, 1935, pp. 262–80 (reprinted as a pamphlet by HMSO, London, 1962, 19 pp.).

CHIDLEY, T. R. E. and KEYS, K. M. (1970) A rapid method for computing areal rainfall. *J. Hydrol.*, **12,** 15–24.

CHIEN, J.-S. and SARIKELLE, S. (1976) Synthetic design hyetograph and rational runoff coefficient. *Proc. Am. Soc. Civ. Engrs., J. Irrig. Drain. Div.*, **102**(IR3), 307–15.

COLLINGE, V. K. (1961) The frequency of heavy rainfalls in the British Isles. *Civ. Engng. Pub. Wks. Rev.*, **56,** 341–4, 497–500.

COURT, A. (1961) Area–depth rainfall formulas. *J. Geophys. Res.*, **66,** 1823–31.

DILLON, E. C. (1954) Analysis of 35-year automatic recordings of rainfall at Cork. *Trans. Instn. Civ. Engrs. Ireland*, **80,** 191–283.

DISKIN, M. H. (1969) Thiessen coefficients by a Monte Carlo procedure. *J. Hydrol.*, **8,** 323–35.

DISKIN, M. H. (1970) On the computer evaluation of Thiessen weights. *J. Hydrol.*, **11,** 69–78.

ESCRITT, L. B. (1950) Design of a network of surface water sewers without recourse to graphical aid (correspondence). *Surveyor*, **109,** 184.

FELGATE, D. G. and READ, D. G. (1975) Correlation analysis of the cellular structure of storms observed by raingauges. *J. Hydrol.*, **24,** 191–200.

FOLLAND, C. K. (1978) Rainfall profiles recommended in Road Note 35. *Chart. Munic. Engr.*, **105,** 169–74.

Frederick, R. H. (1978) Interstorm relations in Pacific Northwest. *Proc. Am. Soc. Civ. Engrs., J. Hydraul. Div.*, **104**(HY12), 1577–86.

Hargreaves, G. H. (1981) Simplified method for rainfall intensities. *Proc. Am. Soc. Civ. Engrs., J. Irrig. Drain. Div.*, **107**(IR3), 281–8.

Hawken, W. H. (1921) An analysis of maximum runoff and rainfall intensity. *Trans. Instn. Engrs. Austr.*, **2,** 193–215.

Hershfield, D. M. and Wilson, W. T. (1957) Generalising of rainfall–intensity–frequency data. Int. Assoc. Sci. Hydrol., Publ. No. 43, pp. 499–506.

Hershfield, D. M., Weiss, L. L. and Wilson, W. T. (1955) Synthesis of rainfall–intensity–frequency regime. *Proc. Am. Soc. Civ. Engrs.*, **81,** separate No. 744, 6 pp.

Holland, D. J. (1964) Rain intensity frequency in Britain. Meteorological Office, Hydrol. Memo. No. 33, 28 pp.

Holland, D. J. (1967) The Cardington rainfall experiment. *Met. Mag.*, **96,** 193–202.

Judson, C. C. (1933) Runoff calculations, a new method. *Proc. Instn. Munic. Co. Engrs.*, **59,** 861–7.

Keifer, C. J. and Chu, H. H. (1957) Synthetic storm pattern for drainage design. *Proc. Am. Soc. Civ. Engrs., J. Hydraul. Div.*, **83**(HY4), 25 pp.

King, M. V. (1967) Storm runoff from urban areas. *Proc. Instn. Civ. Engrs.*, **37,** 43–56.

Lee, P. S., Lynn, P. P. and Shaw, E. M. (1974) Comparison of multiquadric surfaces for the estimation of areal rainfall. *Hydrol. Sci. Bull.*, **19,** 303–17.

Lloyd-Davies, D. E. (1906) The elimination of storm water from sewerage systems. *Min. Proc. Instn. Civ. Engrs.*, **164,** 41–67.

McIllwraith, J. F. (1953) Rainfall intensity–frequency data for New South Wales stations. *J. Instn. Engrs. Austr.*, **25,** 133–9.

MacLean, D. J. (1945) Rainstorm data (correspondence). *Surveyor*, **104,** 34, 58.

Mandeville, A. N. and Rodda, J. C. (1970) A contribution to the objective assessment of areal rainfall amounts. *J. Hydrol. (N.Z.)*, **9,** 281–91.

Marshall, R. J. (1980) The estimation and distribution of storm movement and storm structure, using a correlation analysis technique and raingauge data. *J. Hydrol.*, **48,** 19–39.

Meteorological Office (1968) Appendix to Hydrol. Memo. No. 33, 10 pp.

Ministry of Health, Departmental Committee on Rainfall and Runoff (1930) Rainfall and runoff. *Proc. Instn. Munic. Co. Engrs.*, **56,** 1172–6.

National Water Council and Department of the Environment (1981) *Design and Analysis of Urban Storm Drainage. The Wallingford Procedure*, Vol. 1: *Principles, methods and practice* (National Water Council, London) 173 pp.

Natural Environment Research Council (1975) *Flood Studies Report*, Vol. II: *Meteorological studies* (NERC, London) 81 pp.

Norris, W. H. (1948) Sewer design and the frequency of heavy rain. *Proc. Instn. Munic. Co. Engrs.*, **75,** 349–64.

Ormsby, M. T. M. (1933) Rainfall and runoff calculations. *Proc. Instn. Munic. Co. Engrs.*, **59,** 889–94.

PANDE, B. B. L. and AL-MASHIDANI, G. (1978) A technique for the determination of areal average rainfall. *Hydrol. Sci. Bull.*, **23,** 445–53.

PILGRIM, D. H. and CORDERY, I. (1975) Rainfall temporal patterns for design floods. *Proc. Am. Soc. Civ. Engrs., J. Hydraul. Div.*, **101**(HYl), 81–95.

PILGRIM, D. H., CORDERY, I. and FRENCH, R. (1969) Temporal patterns of design rainfall for Sydney. *Civ. Engng. Trans., Instn. Engrs. Austr.*, **11,** 9–14.

REICH, B. M. (1963) Short duration rainfall intensity estimates and other design aids for regions of sparse data. *J. Hydrol.*, **1,** 3–28.

ROAD RESEARCH LABORATORY (1963) A guide for engineers to the design of storm sewer systems. Road Note No. 35 (HMSO, London) 20 pp.

ROSEVEARE, L. (1930) Runoff as affecting the flow in sewers. *Proc. Instn. Munic. Co. Engrs.*, **56,** 1177–97.

ROSS, C. N. (1921) The calculation of flood discharges by the use of a time contour plan. *Trans. Instn. Engrs. Austr.*, **2,** 85–92.

SARGENT, D. M. (1981) An investigation into the effects of storm movement on the design of urban drainage systems. Part 1. *Pub. Health Engr.*, **9,** 201–7.

SARGENT, D. M. (1982) An investigation into the effects of storm movement on the design of urban drainage systems. Part 2: Probability analysis. *Pub. Health Engr.*, **10,** 111–17.

SHAW, E. M. and LYNN, P. P. (1972) Areal rainfall evaluation using two surface fitting techniques. *Hydrol. Sci. Bull.*, **17,** 419–33.

SHEARMAN, R. J. (1977) The speed and direction of storm rainfall patterns with reference to urban storm sewer design. *Hydrol. Sci. Bull.*, **22,** 421–31.

SIEKER, F. (1977) Simulation of design storms with probable distributions in time and space for storm drainage systems. *Progr. in Wat. Technol.*, **9,** 509–19.

SINGH, V. P. (1976) A rapid method of estimating mean areal rainfall. *Wat. Resour. Bull.*, **12,** 307–15.

THIESSEN, A. H. (1911) Precipitation for large areas. *Mon. Weath. Rev.*, **39,** 1082–4.

TRANSPORT AND ROAD RESEARCH LABORATORY (1976) A guide for engineers to the design of storm sewer systems. Road Note No. 35, 2nd Edn (HMSO, London) 30 pp.

UNITED STATES WEATHER BUREAU (1954) Rainfall intensities for local drainage design in coastal regions of North Africa, longitude 11°W to 14°E. US Dept. of Commerce, Weather Bureau, Hydrologic Services Div., Cooperative Studies Section, Washington, 20 pp. (reprinted 1977).

UNWIN, D. J. (1969) The areal extension of rainfall records: an alternative model. *J. Hydrol.*, **7,** 404–14.

WATKINS, L. H. (1966) Runoff from combined rural and urban areas. Ch. 7 of Thorn, R. B. (ed.), *River engineering and water conservation works* (Butterworths, London) pp. 111–21.

YEN, B. C. and CHOW, V. T. (1969) A laboratory study of surface runoff due to moving rainstorms. *Wat. Resour. Res.*, **5,** 989–1006.

YEN, B. C. and CHOW, V. T. (1980) Design hyetographs for small drainage structures. *Proc. Am. Soc. Civ. Engrs., J. Hydraul. Div.*, **106**(HY6), 1055–76.

# Part II

# FLOOD HYDROLOGY

# 4

# Introduction to Design Flood Estimation

## 4.1. THE FORMATION AND ANALYSIS OF FLOODS

An understanding of the manner in which a catchment area transforms storm rainfall into a discharge hydrograph is a prerequisite to an appreciation of flood estimation methods. The time sequence of events is illustrated by the sketch in Fig. 4.1. When rain falls over a catchment area, the interception storage on vegetation, soil moisture deficits and storage in depressions over the surface of the ground must all be satisfied before measurable runoff begins to reach the network of stream channels. These 'losses' continue throughout the storm, but at a decreasing rate as the available storage capacity is reduced, so that only a proportion of the total rainfall, known as the 'effective rainfall' or 'rainfall excess', finally appears as 'surface' or 'direct' runoff. During the periods between rainstorms, flow in the stream channels is composed almost entirely of discharges from subsurface sources within the catchment, referred to as 'baseflow', which continues at a decreasing rate with time until rainfall occurs. Direct runoff from the storm is then superimposed on the baseflow, raising the total discharge to a peak value from which the flow recesses when rainfall ceases until the runoff is once again wholly derived from baseflow.

The shape of the flood hydrograph from a catchment area is largely determined by:

(1) storm characteristics, such as the distribution in both time and space of rainfall intensities; and
(2) catchment characteristics, such as area, shape, channel and overland slopes, soil types and their distribution, and other geomorphological and geological features.

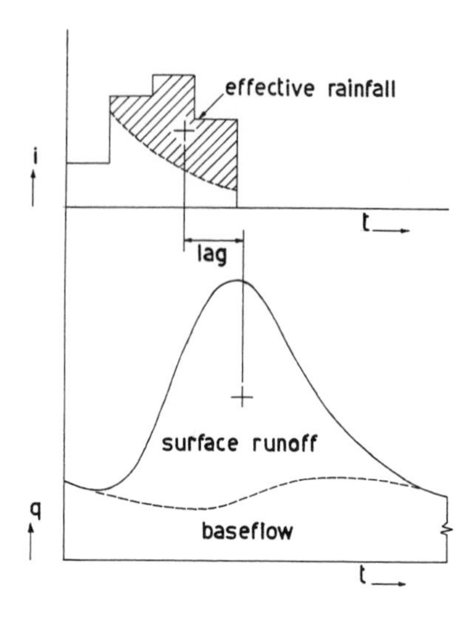


FIG. 4.1. The rainfall–runoff relationship.

Whereas rainfall must be sampled at many points within a catchment area in order to obtain a reasonable estimate of its temporal and spatial variations, streamflow need only be measured at the catchment outfall. River flow data may therefore be considered to represent the integrated effects of both storm and catchment characteristics. Consequently, in order to solve the standard engineering design problem of estimating the flood with a return period of $T$ years at that site, the available streamflow records should be employed to the maximum advantage. In effect, the available approaches fall into one of two categories: (a) statistical techniques; and (b) deterministic methods.

In applying the statistical approach, the analysis may be carried out on either the largest independent peak discharges recorded during each water year (referred to as the annual series), or the largest independent peak flows over a chosen threshold value (known as the partial duration series). In either case, graphical methods or standard statistical techniques, such as the method of moments or the method of maximum likelihood, can be employed to fit a frequency distribution to the series of peak flows. This distribution may then be used to estimate the discharge which corresponds to any specified return period. A comprehensive discussion of the statistical approach to flood estimation is presented in Chapter 5.

In contrast, the deterministic approach to flood estimation is based upon the derivation of a relationship between the discharge and the causative rainfall sequence. In its simplest form, this relationship may consist of an equation for the peak rate of flow in terms of storm and catchment characteristics, as exemplified by the Rational Method:

$$Q = CiA \tag{4.1}$$

where $Q$ is the peak discharge, $A$ is the catchment area, $i$ is the average rainfall intensity during a critical duration referred to as the time of concentration, and $C$ is a coefficient which allows for catchment losses. Despite its simplicity, this 'model' of the rainfall–runoff relationship is sufficient to illustrate the principal characteristic of deterministic flood estimation methods: their dependence upon a knowledge of the magnitude and frequency of the causative events or 'design storms'. The Rational Method is used extensively for the design of small urban drainage systems, an application which is discussed in more detail in Chapter 9. If the complete flood hydrograph is required rather than merely the peak runoff rate provided by eqn (4.1), a more comprehensive description of the rainfall–runoff process is necessary. Of the techniques meeting these requirements, perhaps the most widespread use has been made of the Unit Hydrograph Method introduced by Sherman (1932).

The Unit Hydrograph Method is based upon the premise that direct runoff hydrographs, i.e. streamflow minus the baseflow, caused by effective rainfalls conforming to a standard set of conditions, can be extracted from concurrent records of rainfall and discharge which reflect only the influence of catchment characteristics. Indeed, such direct runoff hydrographs are themselves characteristics of the catchment. Formally, the unit hydrograph of a catchment is defined as the hydrograph of direct runoff resulting from a unit depth (generally 10 mm) of effective rainfall generated uniformly over the area at a constant rate during a specified duration. Once the unit hydrograph has been obtained, it can be employed to construct the hydrograph corresponding to a particular design storm using the principles of proportionality and superposition. Proportionality specifies that the ordinates of the direct runoff hydrograph are proportional to the depth of effective rainfall within the unit duration, whereas superposition ensures that the response to successive blocks of effective rainfall, each occurring within the unit duration, may be obtained by displacing the

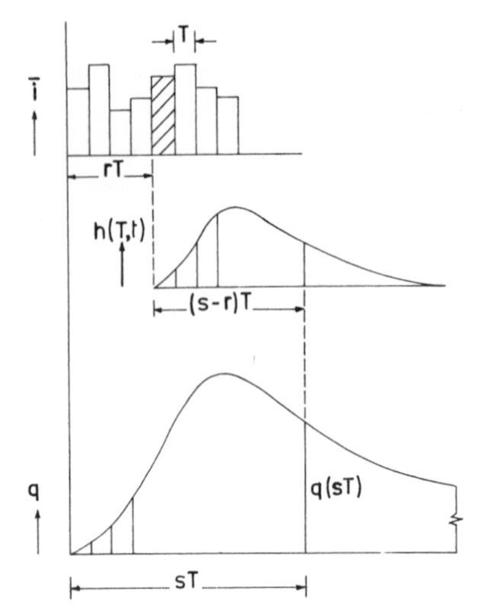


FIG. 4.2. Construction of the direct runoff hydrograph from the effective rainfall distribution and the finite-period unit hydrograph.

individual direct runoff hydrographs and adding their concurrent ordinates. These operations are illustrated in Fig. 4.2, which shows the manner in which the ordinate of the direct runoff hydrograph $s$ time intervals from the beginning of rainfall, $q(sT)$, is formed from the sum of products of the effective rainfall totals in the $r$th time interval and the $(s-r)$th ordinate of the unit hydrograph. More specifically,

$$q(sT) = T \sum_{n=1}^{s} \bar{i}_n h(T, (s-r)T) \tag{4.2}$$

where $h(T, t)$ is the ordinate at time $t$ of the unit hydrograph corresponding to the period $T$, and $\bar{i}_n$ is the average intensity of effective rainfall during the $n$th time interval, $T$.

The function $h(T, t)$ is referred to as the finite-period unit hydrograph or TUH. The principles of proportionality and superposition which underlie its manipulation ensure that the direct runoff is transformed from the effective rainfall by a linear operation, a characteristic which is exploited by the computer-based methods of unit hydrograph derivation discussed in Chapter 6. From a theoretical viewpoint, particular attention has been paid to the function which is obtained when the data interval becomes infinitesimally small, which is known as the instantaneous unit hydrograph or IUH.

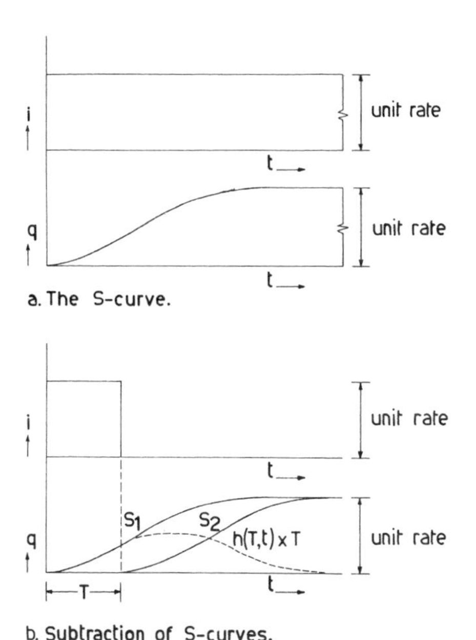


FIG. 4.3. Derivation of the finite-period unit hydrograph using the $S$-curve.

The derivation of the IUH may be readily demonstrated using the so-called $S$-curve, which provides the response of the catchment to an infinitely long unit rate of effective rainfall (see Fig. 4.3(a)). In practice, the $S$-curve is easily constructed from the TUH by superimposing the response to as many unit blocks of effective rainfall as there are TUH ordinates. Given the $S$-curve, the TUH corresponding to any period $T$ (which may differ from that of the TUH from which the $S$-curve was originally derived) may be obtained by subtracting the concurrent ordinates of two $S$-curves displaced by time $T$. As shown in Fig. 4.3(b),

$$h(T, t) = (1/T)(S_t - S_{t-T}) \tag{4.3}$$

As $T$ is allowed to become smaller and smaller, the right-hand side of eqn (4.3) becomes the gradient of the $S$-curve at time $t$. The IUH may therefore be defined by the expression

$$h(0, t) = \mathrm{d}(S_t)/\mathrm{d}t \tag{4.4}$$

Given the IUH instead of the TUH, Fig. 4.2 may be revised as shown in Fig. 4.4. The average rate of effective rainfall at time $\tau$, $i(\tau)$,

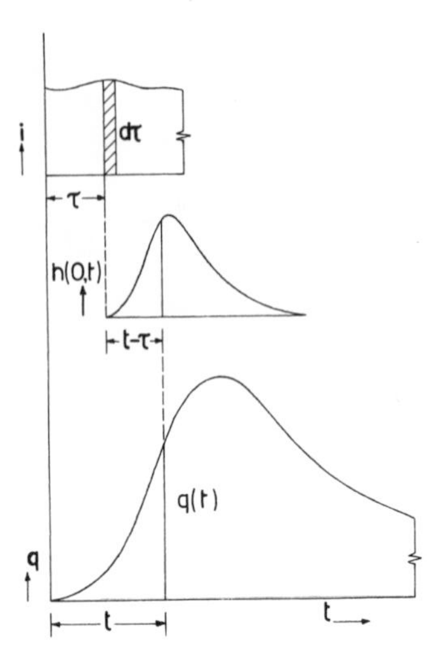


Fig. 4.4. Construction of the direct runoff hydrograph from the effective rainfall distribution and the instantaneous unit hydrograph.

occurring over time $d\tau$ contributes an amount $dq(t)$ to the direct runoff ordinate at time $t$, $q(t)$, where

$$dq(t) = h(0, t-\tau)i(\tau)\, d\tau \tag{4.5}$$

Integrating the contributions from all effective rainfall increments up to time $t$, the direct runoff ordinate $q(t)$ is given by

$$q(t) = \int_0^t h(0, t-\tau)i(\tau)\, d\tau \tag{4.6}$$

Equation (4.6) is known as the convolution integral, and plays a prominent role in some of the theoretical developments relating to unit hydrograph derivation and the simpler linear models of catchment behaviour referred to in Chapter 6.

More elaborate descriptions of the relationship between rainfall and runoff are available in the form of complex models of the land phase of the hydrological cycle. Even with the aid of a digital computer, the fitting of such models, which typically involve the determination of between 10 and 30 parameters, can be tedious and expensive. These models are also considered further in Chapter 6.

The above examples serve to illustrate the range of rainfall–runoff models that is available to form the basis of a deterministic flood estimation method. Unlike the statistical approach, which depends upon the analysis of previous experience, deterministic methods involve the prior selection of the 'design' conditions which lead to the production of the $T$-year flood. The problem of choosing such conditions is further compounded if the catchment of interest is ungauged. Flood estimates must then be obtained by transferring information on catchment response from neighbouring, gauged river basins. This procedure is generally referred to as regionalisation, and forms an important step in the evolution of a flood estimation method whether or not its basis is deterministic or stochastic.

In summary, the derivation of a design flood hydrograph involves wider issues than the application of a rainfall–runoff model. The background to flood estimation methods in general is therefore outlined in this chapter, prior to the more detailed discussion of statistical and deterministic methods in Chapters 5 and 6 respectively. The historical perspective contained in Section 4.2 also presents a framework for choosing the flood estimation method which is the most suitable for solving a particular design problem. The application of regionalisation techniques with both statistical and deterministic flood estimation methods is then considered in Sections 4.3 and 4.4 respectively.

## 4.2. HISTORICAL PERSPECTIVE ON DESIGN FLOOD ESTIMATION

In reviewing the progress made in understanding the relationship between rainfall and runoff since Perrault's study of the Seine river basin between 1668 and 1670, Linsley (1967) divided the intervening period into the following three epochs:

(1) the era of empiricism, extending from 1670 to about 1930, during which the development of flood estimation methods was hampered by lack of data, and heavy reliance was placed on envelope curves, i.e. plots of maximum observed floods against a parameter such as drainage area, or runoff formulae, which provided the same information in the form of an equation;
(2) the era of correlation, covering the quarter-century from 1930 to 1955, which was marked by a growing appreciation of

the factors affecting the transformation of rainfall into runoff; and

(3) the computer era, running from 1955 to the time of writing, during which the capacity to handle raw data and the sophistication of methods of analysis was increased markedly compared with previous practices.

However, in retrospect, flood estimation methods appear to have benefited less from the wider availability of computing facilities than many other aspects of engineering hydrology. In fact, the influence of the era of empiricism and the era of correlation is still very evident in present-day practice. For example Sokolov (1969) drew attention to the continued widespread use of flood formulae in many parts of the world, with due regard being paid to their limits of applicability. The absence of an adequate data base, particularly in developing countries, makes the employment of regional envelope curves inevitable until national hydrometric networks become well established. Even when several years of rainfall and streamflow records are available, the concepts upon which many flood estimation methods are based tend to be borrowed from long-established approaches. The Unit Hydrograph Method (Sherman, 1932), infiltration indices (Cook, 1946) and the antecedent precipitation index (Kohler and Linsley, 1951), all of which were developed during the era of correlation, continue to play a prominent part in current practice.

The longevity of certain techniques is partly explained by the wide spectrum of engineering design problems, ranging from the sizing of spillways for major dams to the spacing of road gulleys, which demand some knowledge of runoff caused by rainfall. The choice of the approach which is the most appropriate for solving a particular design problem is made simpler by recognising that every flood estimation method is limited in its applicability. As listed by Hall (1981), the major factors which should be taken into account when determining the suitability of a method include the following:

(a) *Climate:* Spatial and temporal variations in rainfall intensity tend to differ significantly between different climates. In addition, the balance between precipitation and evaporation affects the antecedent wetness of a catchment, while that between surface water and groundwater influences the magnitude of the baseflow.

(b) *Availability of hydrometric records:* Ungauged catchments require a different approach from that applied to gauged river basins. For drainage areas that have been gauged, further variations in techniques are possible according to both the length of records and the variables being measured.

(c) *The influence of man:* Land use changes, particularly those which alter the form of the drainage network and change the surface properties of the catchment, such as urbanisation, are not generally amenable to treatment by methods intended for application to undisturbed natural areas.

(d) *Size of catchment area:* As the size of the drainage area increases, the relative importance of subcatchment and channel flow processes changes, so that methods developed for large river basins are rarely suitable for small catchment areas (see, for example, Cordery *et al.*, 1981; Pilgrim *et al.*, 1982).

(e) *Type of problem:* In many cases an estimate of the peak flow rate is sufficient for engineering design purposes, but for others, particularly those involving the provision of storage capacity, a complete flood hydrograph may be necessary.

(f) *Design standard:* Differences in approach are often apparent between methods applicable to estimated maximum flood (EMF) conditions and those which deal with floods of a specified return period.

Given the above-mentioned factors, the choice of a suitable flood estimation method may be presented in the form of a flowchart, as shown in Fig. 4.5. Any particular method may apply to the solution of more than one problem. Moreover, in order to ensure that a range of design problems can be covered, several *methods* may be combined into a design *procedure.* This approach is demonstrated very effectively in the United Kingdom Flood Studies Report (Natural Environment Research Council, 1975). The procedure presented in that Report is restricted to the British Isles and does not allow explicitly for the effects of land use changes, but factors (b), (d), (e) and (f) in the above list are all taken into account.

Of the six factors, perhaps the most important is (b), the availability of records. In practice, the situations that can be encountered may range from the densely instrumented catchment with unbroken long-term records to the completely unmonitored river basin. Unfortunately, the majority of flood estimation exercises tend to be concerned with drainage areas that are either ungauged or have a minimal

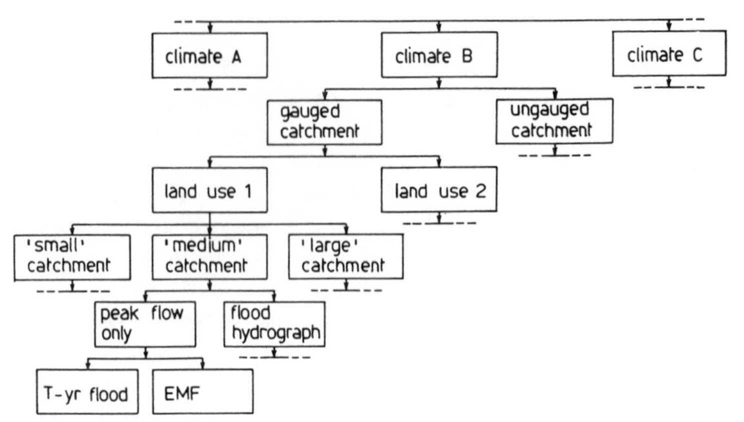


FIG. 4.5. Choice of a suitable flood estimation method (after Hall, 1981, by permission of the Institution of Civil Engineers).

amount of data. As Laurenson has perceptively commented (in discussion to Wolf, 1966), at some stage in the development of a flood estimation method a solution is required to the problem of relating flood hydrograph characteristics to descriptors of the catchment area.

In summarising the results of a review by McCuen *et al.* (1977) on methods of estimating the magnitude and frequency of floods at ungauged sites, McCuen and Rawls (1979) commented that the information provided on different approaches is all too often incomplete. In particular, data requirements are invariably poorly specified, and results are reported in a manner which does not permit a thorough evaluation of the precision and bias in flood estimates, the practicality of the method in terms of both engineer time and computer time and the reproducibility of the results obtained by different individuals. McCuen and Rawls (1979) also suggested that the published literature does not necessarily reflect the extent to which different methods are actually used for design purposes. A more appropriate criterion of status might therefore be the availability of the method in step-by-step form so that additional references are not required in design applications. Furthermore, the place of the method within an overall design procedure, such as that presented in Fig. 4.5, must be clearly established.

## 4.3. REGIONAL FLOOD FREQUENCY ANALYSIS

The concept of relating the magnitude of a design flood to drainage basin characteristics may be seen to date from the derivation of the first runoff formulae. Summaries of such formulae by Jarvis (1936), Richards (1955), Narayana *et al.* (1970), Gray and Wigham (1973) and many others show that the majority assume one of the three following general forms:

$$Q_p = cA^m \tag{4.7}$$

$$Q_p = cA^{nA^{-m}} \tag{4.8}$$

$$Q_p = cA/(a + bA)^m + dA \tag{4.9}$$

where $Q_p$ is the 'maximum' discharge produced by the drainage area $A$, and $a$, $b$, $c$, $d$, $m$ and $n$ are constants and exponents whose values vary from region to region. Equation (4.7) is the simplest and perhaps the most widely quoted of the three, with the exponent $m$ varying from about 0·4 to 0·9. The Craeger curves (Craeger *et al.*, 1945) are probably the best-known examples of eqn (4.8), while eqn (4.9) appears to be typical of continental European practice.

During the first two decades of the twentieth century, the data acquired with the expansion of stream gauging activities were used to advantage by the early practitioners of statistical methods (see Jarvis, 1936). The application of flood magnitude–frequency analysis allowed the illusionary 'maximum' discharge, $Q_p$, of eqns (4.7)–(4.9) to be replaced by $Q_t$, the flood corresponding to a return period of $T$ years. Given the $Q_t$ values for gauged catchments, the prediction of flood flows for ungauged river basins within the same region may be approached in several different ways. For example, separate equations may be derived relating $Q_t$ to catchment characteristics for each return period:

$$Q_t = cA^m B^n \cdots P^r \tag{4.10}$$

where $B, \ldots, P$ are further catchment (or rainfall) characteristics and $r$ is another exponent. The work of Benson (1962) on the floods corresponding to 9 return periods in the north-eastern states of America and of Rodda (1969) on those associated with 4 return periods in England, Wales and Scotland provide ready examples of this approach. Alternatively, one of the most widely applied methods of regionalising flood estimates is that devised by the US Geological Survey and applied during the preparation of numerous state-wide reports during the 1950s.

As described by Dalrymple (1960), the USGS method involves the preparation of two graphs:

(a) a curve showing the variation with return period of the ratio between the $T$-year flood and the mean annual flood, the latter being the average of the floods forming the annual series; and
(b) a plot relating the mean annual flood to the size of the drainage area and other significant catchment characteristics, where appropriate.

This basic approach has been applied in many parts of the world. Where catchment area is adopted as the sole independent variable, the second curve assumes the form of eqn (4.7), with exponents comparable in magnitude to the old flood formulae. The form of the first curve, which is often referred to as the regional 'growth curve', may be expressed graphically following USGS practice (Cole, 1966), or by means of equations similar in form to eqn (4.10) relating the coefficient of variation of observed floods (Nash and Shaw, 1966) or the quotient of the $T$-year flood and the mean annual flood (Gunter, 1974) to catchment characteristics.

The regional flood frequency analysis for the British Isles reported by the Natural Environment Research Council (1975) employed a similar approach to that described by Dalrymple (1960). Initially, Great Britain and Ireland were divided into 11 geographical regions. Owing to a dearth of long streamflow records in one area of southern England, its data were pooled with those of an adjacent region, and 10 regional growth curves were produced. Large differences were evident between some pairs of these curves, particularly at high return periods, which were considered to be acceptable on hydrological grounds and not to represent merely sampling fluctuations about a single national growth curve.

The mean annual flood was expressed as a function of drainage basin characteristics of the general form of eqn (4.10). Such relationships were derived for each of the 11 regions considered in obtaining the growth curves, but a two-stage process of statistical testing was then employed in an attempt to reduce the number of equations. One of the regions in south-east England was found to differ markedly from the rest and was described separately using a three-variable regression equation. Six-variable equations for the other regions were tested firstly for significant differences between the coefficients of individual catchment characteristics. If these coefficients were not found to differ significantly, a second test was carried out for differences in the constant term. This procedure reduced the number of regions from 10 to 6, the equations for which contained the same selection of independent variables with the same coefficients but different values of the constant.

The Flood Studies Report approach may be contrasted with that applied by Benson (1962) in a regional flood frequency analysis of New England. In that study, a plot of the ratio of observed to computed discharges from a regression equation with four independent variables was found to display a consistent geographical variation, which was accounted for by the addition of an 'orographic factor' as a fifth independent variable. In effect, the fit of the derived equation was adjusted by fixing four independent variables and their coefficients and optimising on the extra independent variable.

The apparent success of both of the above-mentioned approaches to deriving a regional equation for the flood of a specified return period points to the possibility that the coefficients for particular combinations of independent variables might be standardised, and the fitting of the equation carried out either by adjusting the constant when residual errors are random or by introducing a derived independent variable when they are spatially variable. The coefficients of regression equations based upon the same independent variables derived from analyses of data from different geographical areas often display a remarkable similarity (see, for example, Natural Environment Research Council, 1975, Vol. 1, Section 4.3.11). However, although such agreement offers some reassurance to the analyst, its generality is open to question on at least two counts.

Firstly, each investigation tends to rely on different methods of both defining and abstracting drainage basin characteristics. The discussion on main channel slope by Benson (1959) readily illustrates the possibilities for concocting different definitions for a single independent variable. In addition, since the detail of a stream network presented on a topographic map is scale-dependent (see Yang and Stall, 1971), methods of measuring the catchment variables can also influence the magnitude of the coefficients in a regression equation. The detailed specification of the procedures which should be applied in abstracting river basin characteristics, such as that presented by the Natural Environment Research Council (1975, Vol. 1, Chap. 4), may at first appear pedantic, but is nevertheless essential for consistency.

Secondly, the strong correlations which exist between certain drainage basin characteristics, such as area and main stream length, is well documented (see, for example, Orsborn, 1974). Inclusion of two such characteristics in a regression analysis as independent variables will also affect the magnitude of the computed coefficients.

## 4.4. REGIONALISATION OF DESIGN FLOOD HYDROGRAPHS

Unlike the statistical approach to flood prediction, which is based upon recorded catchment behaviour, deterministic methods involving the derivation of a design flood hydrograph depend upon the magnitude and frequency of hypothetical causative events or design storms. Such methods require:

(i) the selection of the return period of the design storm that will produce the peak runoff rate of the desired return period;
(ii) the calculation of the depth, duration, areal extent and temporal variations in rainfall intensity of the design storm;
(iii) the determination of the relationship between the total volume of runoff and the total volume of rainfall, and the manner in which the difference between the two (the so-called loss) is distributed in time; and
(iv) the distribution in time of the total volume of runoff in the form of a flood hydrograph.

Until recently, each of the above topics has tended to be treated in relative isolation. Particular attention has been afforded to the modelling of the relationship between rainfall and runoff, a problem which is fundamental to both (iii) and (iv) above. In reviewing such models, a distinction must be drawn between 'single event' models, which are directly applicable to the transformation of a design storm into a flood hydrograph, and 'continuous simulation' models, which produce long synthesised runoff records for subsequent analysis by conventional statistical methods. In both cases the treatment of ungauged river basins poses the same requirement: the correlation of model parameters with catchment characteristics.

As noted in Section 4.1, both single event and continuous simulation models may assume a wide variety of forms, ranging from the linear 'transfer function' methods, such as the unit hydrograph, to multiparameter non-linear models of catchment behaviour An early example of the Unit Hydrograph Method being applied to the ungauged catchment problem may be found in the work of Snyder (1938). This method for synthesising a unit hydrograph is based upon a relationship between lag time, defined as the time interval between the centroid of the effective rainfall distribution and the peak of the direct runoff hydrograph, and two length characteristics of the catchment. These characteristics, with the addition of a term involving main channel slope, have subsequently been incorporated into several design flood estimation techniques (see, for example, Cordery, 1971; Hydrological Research Unit, 1972; US Department of the Interior, Bureau of Reclamation, 1977).

In general, the use of a standard shape of unit hydrograph, particularly the geometrical approximations of the type introduced by the US Department of Agriculture, Soil Conservation Service (1964), has appealed more to practitioners than that of simple linear models which have an analytical form of instantaneous unit hydrograph. The study by Nash (1960) of records from British catchments, in which the parameters of such a model were related to the area, length and slope of a drainage basin, provides a notable exception to this trend. Magette *et al.* (1976) have reported one of the few extensions of this approach to more complex, non-linear models. Those authors found that the values of five parameters obtained from prediction equations produced unacceptable results for two out of five test catchments. In the absence of more encouraging results, the continued popularity of design flood estimation methods based upon well-tried unit hydrograph techniques is perhaps understandable.

When the unit hydrograph approach is applied to flood estimation problems, a design storm must be selected along with a temporal distribution of losses and a baseflow. The design storm is completely specified by the characteristics listed in (ii) above, but of the latter, duration is often selected by a process of trial and error, the 'critical' duration being that which produces the design hydrograph with the largest peak flow rate (see Cordery, 1971; Hydrological Research Unit, 1972; Chow, 1962). In other flood estimation methods based upon the unit hydrograph approach, the response of the drainage area to rainfall is assumed to be sufficiently damped for temporal variations in rainfall intensity, i.e. the storm profile, to be relatively unimportant. Under these somewhat restrictive conditions, Meynink and Cordery (1976) have demonstrated that the 'critical' storm duration may be expressed in analytical form. Those authors also show that relatively small changes in the loss parameters have a marked effect upon the critical storm duration and therefore on the magnitude of the design discharge. Interactions such as this become particularly relevant when considering what is perhaps the most contentious aspect of deterministic flood estimation methods: the relationship between the return period of the peak flow rate and the return period of the causative design storm.

In practice, the return periods of the peak discharge and the design storm are invariably assumed to be equal (Gunter, 1974; Cordery, 1971; Hydrological Research Unit, 1972; US Department of the Interior, Bureau of Reclamation, 1977; US Department of Agriculture, Soil Conservation Service, 1964; Chow, 1962; Fiddes, 1977). This equality is often regarded as being valid for 'average' rather than 'extreme' states of the catchment. Cordery (1971) has suggested as a working hypothesis that the equality may be approximated by setting all other design inputs, i.e. the variables that are significant to the derivation of the design hydrograph which must be specified by the analyst, at their median values. This intuitive argument was substantially justified by the tolerable agreement obtained between flood frequency distributions synthesised by applying the method and those produced by analysing observed floods. However, the median represents only one point on the probability distribution of each design input. A more objective choice of the 'stable' design inputs, i.e. those which on average produce the peak flow rate corresponding to the required return period, may be obtained by means of a sensitivity analysis in which the design method is applied repeatedly with different combinations of the inputs, sampled in proportion to their probabilities of occurrence. This procedure, which is shown schematically in Fig. 4.6, was adopted in formulating the flood estimation method based upon the unit hydrograph technique contained in the United Kingdom Flood Studies Report (Natural Environment Research Council, 1975, Vol. 1, Section 6.7).

As described in the Flood Studies Report and previously by Beran (1974), the four design inputs to the flood estimation method consisted of the total depth and duration of rainfall, the storm profile and a measure of antecedent conditions in the form of a catchment wetness index (CWI). Given four such variables, values may be assigned to three so that the value of the fourth which must be selected to reproduce the required peak discharge can be evaluated. The sensitivity analysis showed that both rainfall depth and CWI were sufficiently flexible to act as the 'free' variable, but neither rainfall duration nor

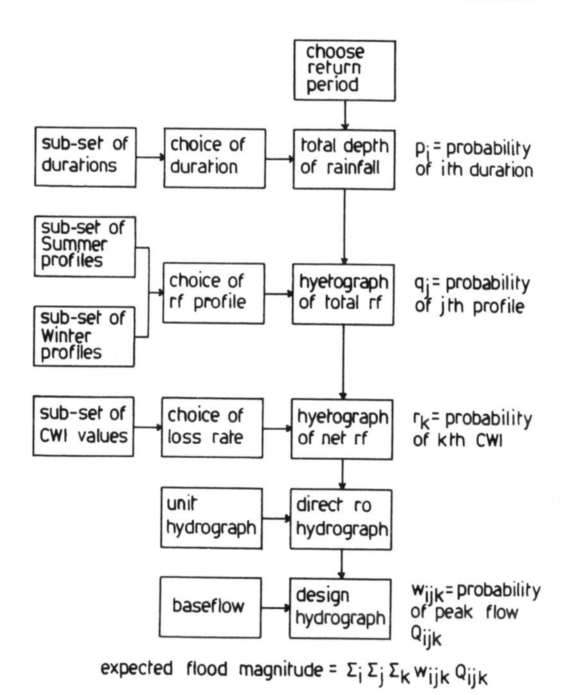


FIG. 4.6. Flowchart for procedure to establish 'stable' design inputs (after Hall, 1981, by permission of the Institution of Civil Engineers).

storm profile proved to be suitable. The rainfall depth could therefore have been specified to ensure an equality between storm and flood return periods, and an appropriate value of CWI determined. However, since this CWI value was found to vary between catchments, the alternative approach of specifying the CWI and constructing a relationship between the return period of the rainfall depth and the required return period of the peak discharge was preferred.

This form of sensitivity analysis is only appropriate where the design inputs can be assumed to be statistically independent. This condition may not be satisfied for the design of urban drainage systems in the more temperate climates, where the severe storm events occur during the summer months when catchments are relatively dry, giving an inverse correlation between rainfall depth and catchment wetness. In these circumstances, Packman and Kidd (1980) have suggested that the synthetic flood frequency distribution should be produced by feeding a long rainfall record through the rainfall–runoff model calibrated to the particular catchment area rather than by means of a sampling experiment. As an example of this approach, those authors describe the procedure for determining the stable design inputs which was applied in the derivation of the Wallingford Procedure for the design of urban drainage systems (see Chapter 9). This model employs the same input variables as the Flood Studies Report unit hydrograph method. However, two of these inputs are fixed externally. The model considers four durations of 15, 30, 60 and 120 min and chooses that which yields the largest flood peak. In addition, the return periods of rainfall depth and peak runoff rate are assumed to be equal, so that the former is chosen automatically. The output from the model was found to be sensitive to storm profile, so that sensible values of an urban catchment wetness index (UCWI) were only obtainable if this remaining rainfall variable was also fixed. The variation in UCWI between catchments was then found to be correlated with variations in average annual rainfall.

This potentially powerful technique for establishing the relationship between the return periods of the flood peak and the causative design storm represents a significant innovation in the development of flood estimation methods. Its advantages over previous approaches to the problem lie principally in the integration of the influence of the significant design inputs in simulating an observed (or synthesised) probability distribution of flows. In effect, the determination of the stable design inputs represents an extension of the model-building phase, and serves to emphasise that the success of a flood estimation method must be judged against a different set of criteria from that of the underlying rainfall–runoff model.

In order to gain acceptance, a new rainfall–runoff model must be superior in its performance to its predecessors. Some predetermined criterion of goodness-of-fit will generally be employed to assess the ability of the model to reproduce observed runoff events. This test should be carried out with storm events that were excluded from the data set on which the model was calibrated, a procedure known as split-record testing, otherwise the exercise merely serves as a check on the arithmetic. Once the model has passed this test, its suitability for design flood estimation must be evaluated according to a supplementary set of criteria.

Firstly, if the model is to be applied to ungauged drainage areas, correlations must be established between model parameters and catchment characteristics, or regional values of the model parameters must be determined. Moreover, the model parameters used as the dependent variables must provide the stable design inputs which ensure that the return period of the peak flow rate bears a specific relationship to the return period of the design storm. Secondly, the limits of application of the flood estimation method based upon the rainfall–runoff model should be clearly defined so that its position in the general flood estimation procedure of Fig. 4.5 may be established. Finally, acceptance of the method into general practice will depend upon the ease with which the technique can be applied, the economy in computation and the reproducibility of the results produced by different analysts.

A cursory examination of many of the more widely used design hydrograph methods, including that in the Flood Studies Report (Natural Environment Research Council, 1975), is sufficient to show their dependence on well-established techniques. Until the additional steps required to develop a rainfall–runoff model into a flood estimation method are more widely appreciated, this apparent reluctance to accept innovation is liable to remain a feature of design practice.

## REFERENCES

Benson, M. A. (1959) Channel-slope factor in flood-frequency analysis. *Proc. Am. Soc. Civ. Engrs., J. Hydraul. Div.*, **85**(HY4), 1–9.

Benson, M. A. (1962) Factors influencing the occurrence of floods in a humid region of diverse terrain. US Geol. Survey, Water-Supply Paper 1580-B, 64 pp.

Beran, M. A. (1974) Estimation of design floods and the problem of equating the probability of rainfall and runoff. In *Design of water resource projects with inadequate data*, Proc. Madrid Symp., UNESCO Studies and Reports in Hydrology No. 16, Vol. 2, pp. 459–76.

Chow, V. T. (1962) Hydrologic design of culverts. *Proc. Am. Soc. Civ. Engrs., J. Hydraul. Div.*, **88**(HY2), 39–55.

Cole, G. (1966) An application of the regional analysis of flood flows. In *River flood hydrology*, Proc. Symp. (Instn. Civ. Engrs., London) pp. 39–57.

Cook, H. L. (1946) The infiltration approach to the calculation of surface runoff. *Trans. Am. Geophys. Un.*, **27,** 726–47.

Cordery, I. (1971) Estimation of design hydrographs for small rural catchments. *J. Hydrol.*, **13,** 263–77.

Cordery, I., Pilgrim, D. H. and Baron, B. C. (1981) Validity of use of small catchment research results for large basins. *Trans. Instn. Engrs. Austr.*, **CE23,** 131–7.

Craeger, W. P., Justin, J. D. and Hinds, J. (1945) *Engineering for dams*, Vol. 1 (Wiley, New York) 246 pp.

DALRYMPLE, T. (1960) Flood-frequency analyses. US Geol. Survey, Water-Supply Paper 1543-A, 80 pp.

FIDDES, D. (1977) Flood estimation for small East African rural catchments. *Proc. Instn. Civ. Engrs., Part 2*, **63,** 21–34.

GRAY, D. M. and WIGHAM, J. S. (1973) Peak flow—rainfall events. Section VIII of Gray, D. M. (ed.), *Principles of hydrology* (Water Information Centre Inc., Port Washington) 24 pp.

GUNTER, B. N. (1974) The investigation of flood estimation procedures for Papua New Guinea. *Proc. Instn. Civ. Engrs., Part 2*, **57,** 635–50.

HALL, M. J. (1981) A historical perspective on the Flood Studies Report. In *Flood Studies Report—Five Years On*, Proc. Symp. (Thomas Telford Ltd, London) pp. 11–16.

HYDROLOGICAL RESEARCH UNIT (1972) Design flood determination in South Africa. Univ. of Witwatersrand, Dept. Civ. Engng., Rept. 1/72.

JARVIS, C. S. (ed.) (1936) Floods in the United States, magnitude and frequency. US Geol. Survey, Water-Supply Paper 771, 497 pp.

KOHLER, M. A. and LINSLEY, R. K. (1951) Predicting the runoff from storm rainfall. US Weather Bureau, Res. Paper 34, 10 pp.

LINSLEY, R. K. (1967) The relation between rainfall and runoff. *J. Hydrol.*, **5,** 297–311.

MCCUEN, R. H. and RAWLS, W. J. (1979) Classification of evaluation of flood flow frequency estimation techniques. *Wat. Resour. Bull.*, **15,** 88–93.

MCCUEN, R. H., RAWLS, W. J., FISHER, G. T. and POWELL, R. L. (1977) Flood flow frequency for ungauged watersheds: a literature evaluation. US Dept. Agric., Agric. Res. Serv., ARS-NE-86, 136 pp.

MAGETTE, W. L., SHANHOLTZ, V. O. and CARR, J. C. (1976) Estimating selected parameters for the Kentucky Watershed Model from watershed characteristics. *Wat. Resour. Res.*, **12,** 472–6.

MEYNINK, C. and CORDERY, I. (1976) Critical duration of rainfall for flood estimation. *Wat. Resour. Res.*, **12,** 1209–14.

NARAYANA, V. V. D., SIAL, M. A., RILEY, J. P. and ISRAELSEN, E. K. (1970) Statistical relationships between storm and urban watershed characteristics. Utah State Univ., Coll. Engng., Wat. Res. Lab., Project Rept. PRWG 74-2.

NASH, J. E. (1960) A unit hydrograph study, with particular reference to British catchments. *Proc. Instn. Civ. Engrs.*, **17,** 249–82.

NASH, J. E. and SHAW, B. L. (1966) Flood frequency as a function of catchment characteristics. In *River flood hydrology*, Proc. Symp. (Instn. Civ. Engrs., London) pp. 115–36.

NATURAL ENVIRONMENT RESEARCH COUNCIL (1975) *Flood Studies Report* (NERC, London) 5 vols.

ORSBORN, J. F. (1974) Determining streamflows from geomorphic parameters. *Proc. Am. Soc. Civ. Engrs., J. Irrig. Drain. Div.*, **100**(IR4), 455–75.

PACKMAN, J. C. and KIDD, C. H. R. (1980) A logical approach to the design storm concept. *Wat. Resour. Res.*, **16,** 994–1000.

PILGRIM, D. H., CORDERY, I. and BARON, B. C. (1982) Effects of catchment size on runoff relationships. *J. Hydrol.*, **58,** 205–21.

RICHARDS, B. D. (1955) *Flood estimation and control*, 3rd Edn (Chapman & Hall, London) 187 pp.

Rodda, J. C. (1969) The significance of characteristics of basin rainfall and morphometry in a study of floods in the United Kingdom. In *Floods and their computation*, Proc. Leningrad Symp., UNESCO Studies and Reports in Hydrology No. 3, Vol. 2, pp. 834–43.

Sherman, L. K. (1932) Streamflow from rainfall by unit-graph method. *Engng. News-Record*, **108,** 501–5.

Snyder, F. F. (1938) Synthetic unit hydrographs. *Trans. Am. Geophys. Un.*, **19,** 447–54.

Sokolov, S. S. (1969) The essence of the problem and the significance of the symposium. In *Floods and their computation*, Proc. Leningrad Symp., UNESCO Studies and Reports in Hydrology No. 3, Vol. 2, pp. 971–80.

United States Department of Agriculture, Soil Conservation Service (1964) *National Engineering Handbook*, Section 4: *Hydrology* (Soil Conservation Service, Washington, DC) 650 pp.

United States Department of the Interior, Bureau of Reclamation (1977) *Design of small dams*, 2nd Edn (US Government Printing Office, Washington, DC) 816 pp.

Wolf, P. O. (1966) Comparison of methods of flood estimation. In *River flood hydrology*, Proc. Symp. (Instn. Civ. Engrs., London) pp. 1–23.

Yang, C. T. and Stall, J. B. (1971) Note on the map scale effect in the study of stream morphology. *Wat. Resour. Res.*, **7,** 709–12.

# 5

# Flood Magnitude–Frequency Analysis

## 5.1. INTRODUCTION

Perhaps the most common problem encountered by an engineering hydrologist is the estimation of the flood discharge corresponding to a specified return period or recurrence interval, $T$, at a particular point on a river sytem. The magnitude of this discharge, which is often referred to as the $T$-year flood, is that which can be expected to be exceeded once on average every $T$ years. The estimation of this discharge requires a knowledge of the statistical distribution of floods at the site of interest. Any observations of floods which might be available at that site constitute a sample from a statistical population. The basic problem of flood frequency analysis is therefore the inference of the form and parameters of the statistical distribution which characterises the population from the sample of recorded data.

In the majority of cases the observed floods are composed of the sequence of maximum instantaneous discharges recorded during successive water years. This so-called annual flood series may be contrasted with the partial duration series, which consists of all independent flood peaks within the record that exceed a specified threshold discharge. When an annual flood series is analysed, the probability of the $T$-year flood being exceeded in any single water year is $1/T$. The cumulative distribution function (CDF) describing the annual floods, $F(q)$, i.e. the probability that an annual flood drawn at random from the population is less than or equal to $q$, is therefore related to return period by the expression

$$F(q) = 1 - (1/T) \tag{5.1}$$

Although design floods are almost always specified by their return period, a far more pertinent label would be the probability that the chosen discharge would be exceeded during the design life of the project. This probability, which is also referred to as the simple risk of failure, *PF*, may be developed directly from eqn (5.1). If the design life is taken as $N$ years, then the probability of non-exceedance in $N$ years is given by $F(q)$ raised to the power $N$. The probability of exceedance in $N$ years is therefore given by

$$PF = 1 - [1 - (1/T)]^N \tag{5.2}$$

Graphical solutions to this equation have been presented by Yen (1970) among others. Equation (5.2) shows that if the design flood has a return period of 50 years, the risk of failure in a 50-year design life is 0·64; if the design life were 25 years, the risk would be approximately 0·4. This approach is particularly relevant to the design of temporary works, such as coffer dams (see Cochrane, 1967), but is capable of application in a wider context. A risk-based approach has been applied to storm sewer design by Yen *et al.* (1974), to flood levee systems by Tung and Mays (1981a, b) and to bridge waterway openings by Tung and Mays (1982), but this more rigorous methodology has yet to achieve a wider following.

The occasional user of statistical methods for flood frequency analysis is often inhibited by unfamiliar terminology and the variety of methods of notation favoured by different authors. Section 5.2 therefore discusses the various forms in which a frequency distribution can be presented, along with the statistical moments that are employed to characterise its shape and the forms of estimator that are used to compute these moments from finite samples of data. The different methods that are available for fitting a frequency distribution to an observed sample of annual floods are then described in Section 5.3.

In general, annual flood series are found to have a frequency distribution which is both unimodal, i.e. having a single peak, and positively skewed, i.e. having a longer tail to the right of the peak than to the left. Such behaviour may be described by statistical distributions having only two parameters, although those having three are capable of providing a greater flexibility in the fitting process. The option is also available to apply the same two- and three-parameter distributions to describe transformations of the original observations, such as their logarithms or their square roots. Recently, a more complex form of distribution having five parameters, referred to as the Wakeby distribution, has been introduced (see Houghton, 1978a,b; Landwehr *et al.*, 1979b,c). Five parameters are also required if a mixture of two two-parameter distributions is employed, such as that of two normal distributions considered by Singh (1968). However, since neither of these options has as yet been used extensively for flood frequency analysis, the discussion of statistical distributions in Section 5.4 is confined to the more widely used alternatives.

The fitting of frequency distributions to annual flood series is likely to yield unreliable estimates of the $T$-year flood when $T$ is in excess of the record length unless 10 years or more of data are available. For shorter periods of records, a partial duration series analysis, based upon the 'peaks over a threshold' (POT) model, is to be preferred. The details of this approach are therefore presented in Section 5.5. If the site of interest is ungauged, then recourse must be made to regional equations of the type already discussed in Section 4.3 in order to provide estimates of the $T$-year flood.

Having selected an appropriate statistical distribution to describe a series of observed floods and estimated the parameters of that distribution from the sample of data, a method is required for assessing the success with which the chosen distribution describes the record. Goodness-of-fit indices, such as the chi-square or the Kolmogorov–Smirnov index, are available for this purpose, and are described briefly in Section 5.6. Unfortunately, as Matalas and Wallis (1973) among others have pointed out, such tests are often not powerful enough to discriminate between reasonable choices of distribution. Slack *et al.* (1975) have therefore suggested that attention should be concentrated on choosing a distribution to estimate the $T$-year flood that minimises any economic losses associated with either underdesign or overdesign. This approach would involve the specification of the shape of the loss function as well as the choice of the underlying distribution of floods. In the absence of information on either, Slack *et al.* (1975) showed by means of an extensive random sampling experiment that the normal distribution would be the preferred selection. Supplementary work by Wallis *et al.* (1976) indicated that the choice of statistical distribution was not overly sensitive to the number of the available observations.

In the absence of objective criteria for choosing between different statistical distributions to describe an annual flood series, an attempt has been made in the United States to standardise upon the use of the log Pearson type 3 distribution by Federal Agencies (see Benson, 1968). In the United Kingdom an implicit preference has been expressed for the general extreme value distribution (see Natural Environment Research Council, 1975). Some insight into the appropriateness of such choices may be gained from the results of further random sampling experiments, which have been reported by Matalas *et al.* (1975), Wallis *et al.* (1977) and Landwehr *et al.* (1978, 1980) and are summarised briefly in Section 5.6.

## 5.2. FREQUENCY DISTRIBUTIONS, MOMENTS AND STATISTICAL ESTIMATORS

The frequency distribution which characterises a random variable, $x$, may be described in terms of its cumulative distribution function (CDF), $F(x)$, which gives the probability of non-exceedance of $x$, or its probability density function (PDF), $f(x)$, which is defined by the derivative of the CDF:

$$f(x) = \mathrm{d}F(x)/\mathrm{d}x \tag{5.3}$$

Alternatively, the random variable $x$ may be expressed as a function of a second variable, $y$, whose CDF, $F(y)$, is known. A typical form of relationship is

$$x = c + ay \tag{5.4}$$

where $c$ and $a$ are parameters of the distribution of $x$, and for the distribution of $y$ the equivalent parameters are $c = 0$ and $a = 1$. The variable $y$ is referred to as the standardised or reduced variate with respect to $x$.

The latter method of presentation is particularly relevant to graphical approaches to distribution fitting. Whereas $x$ is a non-linear function of the probability $F(x)$ when the latter is plotted on a linear scale, as shown in Fig. 5.1(a), the corresponding relationship between $x$ and $y$ is linear when arithmetic scales are used, as depicted in Fig. 5.1(b). The use of standardised variates therefore obviates the need for supplies of the special graph papers for different distributions that have abscissae which linearise each CDF. However, although standardised variates are employed extensively for manipulating observations, the CDF remains the more convenient form from a theoretical standpoint.

A frequency distribution may be characterised in terms of its moments, which have an analogous definition to that of the moments of area and mass that are used in mechanics. More formally, if $x_0$ is a specified value of a random variable, $x$, having a PDF, $f(x)$, then the

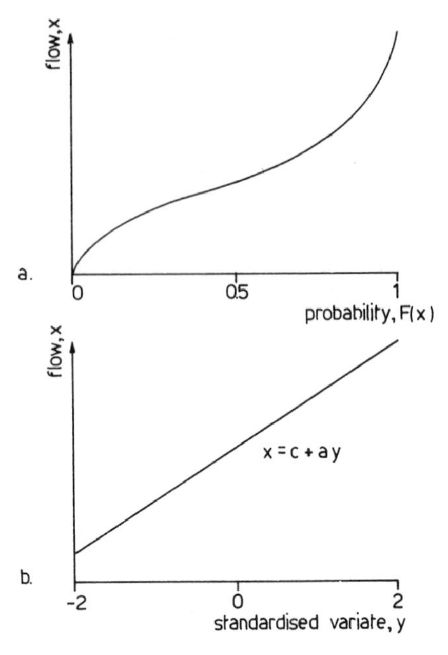


FIG. 5.1. Representation of flood frequency distribution.

$m$th moment about $x_0$ is given by

$$\mu'_m = \int_{-\infty}^{\infty} (x - x_0)^m f(x)\,dx \tag{5.5}$$

When $x_0 = 0$, eqn (5.5) gives the $m$th moment about the origin, which for $m = 1$ defines the mean or expected value, $\mu$, of the variable $x$. The second and higher moments are conventionally taken about the mean, and referred to as the central moments, $\mu_m$. The second central moment, also referred to as the variance, $\sigma^2$, is given by

$$\mu_2 = \sigma^2 = \int_{-\infty}^{\infty} (x - \mu)^2 f(x)\,dx \tag{5.6}$$

The square root of the variance, $\sigma$, which is referred to as the standard deviation, provides a measure of the spread of the values of $x$ and is therefore a parameter of scale. The coefficient of variation, which is defined by the quotient of the standard deviation and the mean, $\sigma/\mu$, is frequently employed as a dimensionless measure of dispersion.

The third central moment, $\mu_3$, provides a measure of asymmetry of the frequency distribution of $x$, and is generally incorporated into another dimensionless ratio known as the skewness, which is given by

$$\gamma = \mu_3/\sigma^3 \tag{5.7}$$

When $\gamma$ is positive (negative), the right-hand tail of the distribution is larger (smaller) than the left-hand tail. The closer the value of $\gamma$ to zero, the more symmetrical the frequency distribution.

The fourth central moment, $\mu_4$, is also generally expressed in terms of a dimensionless ratio referred to as the kurtosis:

$$\lambda = \mu_4/\sigma^4 \tag{5.8}$$

which provides a measure of the flatness of the peak of the frequency distribution.

The above-mentioned definitions of $\mu, \sigma, \gamma$ and $\lambda$ refer to the population of random variables, $x$. In practice, analysis is confined to random samples of $n$ values taken from the population, $x_i$, $i = 1, 2, \ldots, n$. The arithmetic mean of these values is given by

$$\bar{x} = (1/n) \sum x_i \tag{5.9}$$

where the summation is taken over all $n$ values of $x_i$. $\bar{x}$ is a sample estimate of the population mean, $\mu$. Other random samples of $n$ values from the same population will generally yield different values of $\bar{x}$. Just as the $x_i$ have a statistical distribution, so the values of $\bar{x}$ computed from independent samples of size $n$ can be described by a function, known as the sampling distribution, whose form can be deduced from that of the $x_i$. Similarly, all other sample estimates, such as the standard deviation, skewness, and largest and smallest values, have their own sampling distribution. All such estimates are referred to as the statistics of the distribution, and may have different properties according to the form of the equation by which they are computed. For example, the sample estimate of the variance may be calculated from the average of the squares of the differences of each observation from the mean according to the equation

$$s^2 = (1/n) \sum (x_i - \bar{x})^2 \tag{5.10}$$

where the summation is again taken over all $n$ values of $x_i$. Unfortunately this estimator is biased, i.e. when the values of $s^2$ from many samples are averaged, they consistently underestimate the population value of $\sigma^2$. For this reason, the estimator

$$s^2 = (n-1)^{-1} \sum (x_i - \bar{x})^2 \tag{5.11}$$

is almost always preferred to that of eqn (5.10). Strictly, the magnitude of the bias correction, $\sigma^2 - s^2$, is also dependent upon the form of the statistical distribution of the $x_i$.

Similarly, the coefficient of skewness, g, may be estimated from samples of size $n$ according to the equation

$$g = [(1/n) \sum (x_i - \bar{x})^3]/s^{1\cdot 5} \tag{5.12}$$

or the expression

$$g = [\{n/(n-1)(n-2)\} \sum (x_i - \bar{x})^3]/s^{1\cdot 5} \tag{5.13}$$

Both forms of estimator are both biased and bounded algebraically (Kirby, 1974) such that

$$-(n-2)/\sqrt{(n-1)} \leq g \leq (n-2)/\sqrt{(n-1)} \tag{5.14}$$

This condition applies irrespective of the frequency distribution of the $x_i$ and the population value of $\gamma$. Bias correction factors for both the standard deviation and the coefficient of skewness obtained by means of extensive random sampling experiments have been presented by Wallis *et al.* (1974). These results have been employed by Bobee and Robitaille (1975) to derive analytical expressions for the bias correction factors for three common statistical distributions with skewness computed according to eqn (5.12).

The values of $\bar{x}$ calculated according to eqn (5.9) provide unbiased estimates of the population mean, $\mu$.

## 5.3. FITTING PROCEDURES

### 5.3.1. Graphical Methods

When graphical fitting methods are applied, the flood discharge is regarded as a function of a standardised variate with a known statistical distribution. The data sample is plotted as a series of $n$ discrete points, and a straight line is fitted to the sample to give an estimate of the population relationship between flood flows and the standardised variate. The use of a straight line as opposed to a curve is a subjective decision on the part of the analyst. Nevertheless, such is the sampling variability of the larger floods in particular that, with only a small number of observations, the refinement provided by a curvilinear fit is hardly justified.

The $n$ coordinates that are required for graphical estimation are the ordered set of sampled data, $x(1) \leqslant x(2) \leqslant \cdots x(i) \cdots \leqslant x(n)$, and their corresponding standardised variates, $y(1)$, $y(2), \ldots, y(i), \ldots, y(n)$, which are referred to as their plotting positions and may also be specified as a probability, $F(i)$. Bearing in mind that the $i$th smallest observation in an ordered sample, $x(i)$, is itself a random variable, the $y(i)$ or $F(i)$ should ideally be chosen so that the mean of $x(i)$ when plotted lies on the population line (see Cunnane, 1978). This plotting position depends upon the sample size, $n$, as well as the form of the statistical distribution. Many plotting position formulae are special cases of the general expression

$$F(i) = (i - \alpha)/(n + 1 - 2\alpha) \tag{5.15}$$

where for the normal distribution $\alpha = 3/8$ (the Blom formula) and for the extreme value type 1 distribution $\alpha = 0{\cdot}44$ (the Gringorten formula). Cunnane (1978) has recommended $\alpha = 0{\cdot}4$ as a compromise value, and drawn attention to the bias leading to overestimation at high return periods associated with some of the more widely used expressions, such as the Weibull formula for which $\alpha = 0$.

### 5.3.2. Method of Moments

The method of moments is based upon the principle that a frequency distribution is uniquely defined if all its moments are known. For the majority of distributions in common use, the first two or perhaps three moments provide an adequate specification. Moreover, the number of moments required is equal to the number of distribution parameters. The location parameter is analogous to the first moment about the origin, the scale parameter to the square root of the second central moment, and the shape parameter if present depends upon the third central moment. Basically, therefore, the method of moments depends upon the development of relationships between the moments and the parameters of each distribution.

### 5.3.3. Method of Maximum Likelihood

Given a PDF, $f(x; a, b, \ldots)$ of a random variable, $x$, with parameters, $a, b, \ldots$, which are to be estimated, the expression

$$L = \Pi f(x; a, b, \ldots) \tag{5.16}$$

where the symbol $\Pi$ represents the product of all values of the PDF corresponding to the $n$ sampled observations, is called the likelihood function. In the method of maximum likelihood, the parameters, $a, b, \ldots$, are estimated so that $L$ attains its maximum value. For many distributions, $\ln L$, which is a monotonically increasing function of $L$, referred to as the log likelihood function, is employed in preference to $L$ for ease of computation. The likelihood equation may therefore be written as

$$\ln L = \ln \Pi f(x; a, b, \ldots) = \sum \ln f(x; a, b, \ldots) \qquad (5.17)$$

where the summation is taken over all $n$ sampled values. Estimates of the parameters are obtained by taking the partial derivatives of $\ln L$ with respect to each and equating the results to zero:

$$(\partial \ln L)/(\partial a) = 0; \qquad (\partial \ln L)/(\partial b) = 0; \ldots \qquad (5.18)$$

Parameter estimates computed by the method of maximum likelihood are usually unbiased and are generally regarded as the most efficient obtainable. However, eqns (5.18) give rise to a group of non-linear expressions for certain distributions, and their solution can be both difficult to program and time-consuming in operation. Moreover, a solution may prove to be unobtainable for certain data sets.

### 5.3.4. Other Fitting Methods

The three methods outlined above are generally the most widely employed in fitting frequency distributions to hydrological data. However, alternative techniques have been developed for individual distributions or types of distribution, which have been shown to possess significant advantages over the more conventional approaches. For example, for distributions that can only be expressed in what is known as an inverse form, $x = x(F)$, rather than the more familiar $F = F(x)$, or for distributions that can be defined in both forms, Landwehr *et al.* (1979a) have advocated the method of probability weighted moments as described by Greenwood *et al.* (1979). Techniques that have been developed for specific distributions which are worthy of further consideration are referred to in Section 5.4.

## 5.4. STATISTICAL DISTRIBUTIONS

### 5.4.1. The Normal Distribution

The PDF of the normal or Gaussian distribution is given by

$$f(x) = [\sqrt{(2\pi)}a]^{-1} \exp\{-[(x-c)/a]^2/2\}; \qquad -\infty < x < \infty \qquad (5.19)$$

and the CDF by the integral

$$F(x)=\int_{-\infty}^{x} f(x)\,\mathrm{d}x \tag{5.20}$$

$F(x)$ may be computed numerically for given values of the location and scale parameters, $c$ and $a$, which for this distribution are also the mean, $\mu$, and standard deviation, $\sigma$, respectively. The standardised normal variate is defined as

$$y=(x-\mu)/\sigma \tag{5.21}$$

and its CDF by

$$G(y)=\int_{-\infty}^{y} \sqrt{(2\pi)^{-1}}\exp(-y^2/2)\,\mathrm{d}y \tag{5.22}$$

Since eqn (5.22) contains no unknown parameters, $G(y)$ may be tabulated simply as a function of y and is available in the majority of statistical textbooks and sets of tables, such as Lindley and Miller (1961).

Although the normal distribution has the advantages of simplicity and convenience, its utility for flood frequency analysis is limited by its symmetry. However, the pronounced skewness of many annual flood series may be reduced considerably by the application of an appropriate transformation. The family of such transformations suggested by Box and Cox (1964) may be written as

$$\begin{aligned} z &= (x^K-1)/K; \qquad K\neq 0 \\ z &= \ln(x); \qquad K=0 \end{aligned} \tag{5.23}$$

where $K$ is a constant whose value is chosen to minimise the skewness of $z$. Hinkley (1977) has suggested that in order to obtain an appropriate $K$-value, the sample mean, standard deviation and median (the value exceeded in magnitude by half the data set) of the transformed data, $\bar{z}$, $s_z$ and $z_m$, should be computed for $K=0, \pm 0{\cdot}5, \pm 1{\cdot}0, \ldots$. The $K$ which minimises the quantity $(\bar{z}-z_m)/s_z$ is then found by interpolation. Alternatively, a maximum likelihood estimator of $K$ has been presented by Chander *et al.* (1978).

Another transformation for normalising skewed frequency distributions has been suggested by Bethlahmy (1977). Referred to as the SMEMAX (SMall, MEdium, MAXimum) transformation, this approach is based upon the trigonometrical solution of a right-angled

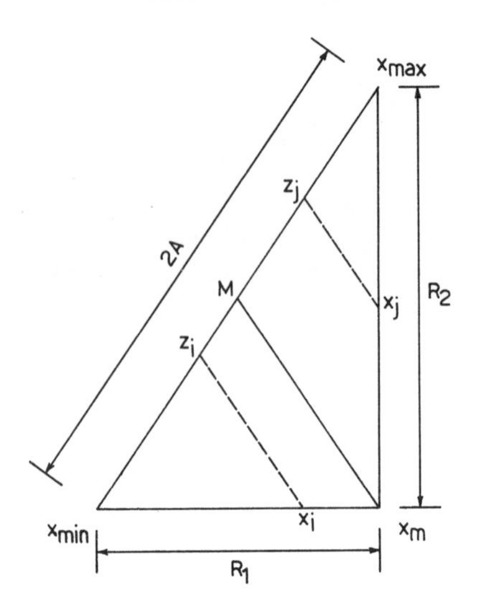


FIG. 5.2. The SMEMAX transformation.

triangle, as shown in Fig. 5.2. The lengths of the two sides of the triangle are given by

$$R_1 = x_m - x_{min}; \qquad R_2 = x_{max} - x_m \tag{5.24}$$

where $x_{min}$, $x_m$ and $x_{max}$ are the minimum, median and maximum values within the sample of data. The point $M$ (see Fig. 5.2) bisects the hypotenuse, whose length is $2A$. The points $z_i$ and $z_j$ on the hypotenuse are the transformed values of the original observations, $x_i$ and $x_j$, where $x_iz_i$ and $x_jz_j$ are drawn parallel to the line $Mx_m$. As demonstrated by Venugopal (1980), the transformed values, as measured from the vertex $x_{min}$, may then be obtained directly from the geometry of the triangle as

$$\begin{aligned} z_i &= (x_i - x_{min})(A/R_1); \qquad & x_i \leqslant x_m \\ z_i &= A + (x_i - x_m)(A/R_2); \qquad & x_i > x_m \end{aligned} \tag{5.25}$$

### 5.4.2. The Log-normal Distribution

If the random variable $x$ follows the three-parameter log-normal distribution, then the variable $z = \ln(x - c)$ is normally distributed with a mean $\mu_z$ and a variance $\sigma_z^2$. Its PDF may therefore be written as

$$f(x) = [(x-c)\sigma_z\sqrt{(2\pi)}]^{-1}\exp\{-[(\ln(x-c) - \mu_z)/\sigma_z]^2/2\} \tag{5.26}$$

The mean, variance and skewness of the variable $x$ are related to the mean and variance of the transformed variable $z$ and the parameter $c$ by the equations

$$\mu_x = c + \exp(\mu_z + \sigma_z^2/2) \tag{5.27}$$

$$\sigma_x^2 = [\exp(\sigma_z^2) - 1] \exp(2\mu_z + \sigma_z^2) \tag{5.28}$$

$$\gamma_x = [\exp(3\sigma_z^2) - 3\exp(\sigma_z^2) + 2]/[\exp(\sigma_z^2) - 1^{1 \cdot 5}] \tag{5.29}$$

A direct method of solving eqns (5.27)–(5.29) for $\mu_z$, $\sigma_z^2$ and $c$ has been outlined by Charbeneau (1978). An alternative approach using the median instead of the coefficient of skewness was presented by Sangal and Biswas (1970), but Burges *et al.* (1975) have demonstrated that estimates of $c$ obtained using this technique are both more variable and more biased than those based upon the skewness.

When the parameter $c$ is set to zero, eqn (5.26) reduces to the form of a two-parameter log-normal distribution. For this distribution, the coefficient of variation of $x$, CV, can be simply expressed as

$$\mathrm{CV} = \sqrt{[\exp(\sigma_z^2) - 1]} \tag{5.30}$$

and the coefficient of skewness is related to CV by

$$\gamma_x = 3\mathrm{CV} + \mathrm{CV}^3 \tag{5.31}$$

The details of several alternative fitting methods for both the two-parameter and the three-parameter distributions have been discussed by Stedinger (1980). Whichever technique is adopted, probability calculations are more conveniently carried out in terms of the normal variates having a mean $\mu_z$ and variance $\sigma_z^2$. The reverse transformation

$$x = c + \exp(z) \tag{5.32}$$

may then be applied.

### 5.4.3. The Pearson Type 3 Distribution

The PDF of the Pearson type 3 distribution may be written as

$$f(x) = [a^b \Gamma(b)]^{-1} (x - c)^{b-1} \exp\{-(x - c)/a\} \tag{5.33}$$

where $a$, $b$ and $c$ are the scale, shape and location parameters of the distribution respectively, and $\Gamma$ is the complete gamma function. The CDF is given by

$$F(x) = \int_c^x f(x)\,\mathrm{d}x \tag{5.34}$$

The mean, $\mu$, variance, $\sigma^2$, and skewness, $\gamma$, are related to the distribution parameters by the equations

$$\mu = c + ab \tag{5.35}$$

$$\sigma^2 = a^2 b \tag{5.36}$$

$$\gamma = 2/\sqrt{b} \tag{5.37}$$

When $\gamma$ is positive, $a$ is positive and $c < x < \infty$; when $\gamma$ is negative, $a$ is negative and $-\infty < x < c$, i.e. $c$ becomes an upper bound or 'maximum' flood. The standardised variate of the distribution is

$$y = (x - c)/a \tag{5.38}$$

A special case of eqn (5.33) occurs when $c = 0$, yielding the PDF of the gamma distribution:

$$f(x) = [a^b \Gamma(b)]^{-1} x^{b-1} \exp(-x/a); \qquad b > 0 \tag{5.39}$$

Its CDF is given by

$$F(x) = \int_0^x f(x)\,dx \tag{5.40}$$

which is also known as the incomplete gamma function and can be evaluated by numerical integration. Equations (5.35)–(5.37) with $c$ set to zero also apply to the gamma distribution. Similarly, the standardised variate is given by the quotient $x/a$, and its CDF may be written as

$$G(y) = \int_0^y \{\Gamma(b)\}^{-1} y^{b-1} \exp(-y)\,dy \tag{5.41}$$

Since eqn (5.41) contains the parameter $b$, each positive value of the latter describes a different PDF and tabulations of $G(y)$ require a separate one-way table for each $b$. When $b$ becomes very large, the skewness becomes very small and the distribution approaches the normal form.

Another special case of particular interest is that in which $b$ is unity. Equation (5.33) then reduces to the PDF of the exponential distribution:

$$f(x) = a^{-1} \exp\{-(x - c)/a\} \tag{5.42}$$

Equations (5.35)–(5.37) with $b = 1$ are applicable to the exponential distribution. The standardised variate is given by eqn (5.38), and eqn (5.41) for its CDF reduces to

$$G(y) = 1 - \exp(-y) \tag{5.43}$$

Since this expression contains no parameters, $G(y)$ may be set out in a simple one-way table.

Although Bobee and Robitaille (1977) in a comparative study found that the Pearson type 3 distribution conformed better in general with annual flood series than the log Pearson type 3 (see Section 5.4.4), the former does not appear to have achieved the popularity of the latter. Indeed, the Pearson type 3 distribution was not even included in the work reported by Benson (1968), which recommended the log Pearson type 3 distribution as the base method of flood frequency analysis to be used by all US Federal agencies. However, Buckett and Oliver (1977) have advised that particular care is needed in applying the Pearson type 3 distribution towards the lower end of its range owing to the radically different shapes that result for different $b$-values. Those authors suggest that, for the majority of hydrological applications, $b$ should equal or exceed zero and preferably exceed unity, but note that values of this parameter are extremely sensitive to the fitting procedure which is employed. Matalas and Wallis (1973) have presented a maximum likelihood method for the Pearson type 3 distribution, but warn that, for small sample sizes, solutions may prove to be impossible because of computer-imposed restraints. A similar result was obtained by Bobee and Robitaille (1977), who also recommended that the sample skewness should be corrected for bias in applying the method of moments as set out in eqns (5.35)–(5.37).

### 5.4.4. The Log Pearson Type 3 Distribution

If the random variable $x$ follows the log Pearson type 3 distribution, then $z = \ln(x)$ conforms to the Pearson type 3 distribution. The PDF of $x$ is given by

$$f(x) = [a^b \Gamma(b)]^{-1}\{\ln(x) - c\}^{b-1} x^{-(1+a)/a} \exp(c/a) \qquad (5.44)$$

where the parameters $a$, $b$ and $c$ are common to both the Pearson type 3 and the log Pearson type 3 forms of distribution. For the latter, the mean, variance and skewness of $x$ are related to $a$, $b$ and $c$ by the following equations:

$$\mu = \exp(c)[1-a]^{-b} \qquad (5.45)$$

$$\sigma^2 = \exp(2c)[(1-2a)^{-b} - (1-a)^{-2b}] \qquad (5.46)$$

$$\gamma = [\exp(3c)/\sigma^3][(1-3a)^{-b} - 3\{(1-2a)(1-a)\}^{-b} + 2(1-a)^{-3b}] \qquad (5.47)$$

As noted by Rao (1980a), eqns (5.45)–(5.47) are difficult to solve for the parameters. Moreover, when $(1/a)>0$, the moments of the log Pearson type 3 distribution $\geqslant(1/a)$ cannot be defined. The overall geometrical shape of the distribution is governed by the scale and shape parameters, $a$ and $b$. Rao (1980a) has listed the possibilities as follows:

| | | |
|---|---|---|
| $(1/a)>0$: | $0<b\leqslant 1$ | reverse J-shape |
| | $b>1$ | unimodal (skewed) bell-shape |
| $(1/a)<0$: | $-1<(1/a)<0; 0<b\leqslant 1$ | U-shape |
| | $(1/a)<-1; 0<b\leqslant 1$ | J-shape |
| | $(1/a)<-1; b>1$ | unimodal bell-shape |

According to Rao (1980a), the boundary between the two classes $(1/a)>0$ and $(1/a)<0$ corresponds to the form of the two-parameter log-normal distribution. The hydrologically unlikely U- and J-shapes occur as the log Pearson type 3 distribution departs more and more from the two-parameter log-normal.

A wide variety of methods is available for estimating the parameters of the log Pearson type 3 distribution. The approach recommended in the study reported by Benson (1968) involved applying the method of moments to the logarithms of the observed data, i.e. fitting a Pearson type 3 distribution to the transformed observations. An alternative technique presented by Bobee (1975), in which the method of moments is applied to the data as recorded to obtain the parameters, and the latter are then inserted in eqns (5.35)–(5.37) to obtain the statistics of the transformed observations, has the advantage of preserving the moments of the original data. These techniques, along with a maximum likelihood method proposed by Condie (1977), were included in a sampling experiment by Nozdryn-Plotnicki and Watt (1979). Those authors showed that no one method was superior to the others in computing estimates of the $T$-year flood for all feasible parameter values. However, the method proposed by Bobee (1975) performed better when the skewness of the transformed data was negative, and the method of moments was superior for positive skewness coefficients.

The principal disadvantage to the conventional method of moments is the need to employ the skewness, which is known to be a statistic subject to bias and a large sampling error. Rao (1980b) has therefore suggested that advantage should be taken of the common parameters in the transformed and untransformed distributions to employ an estimation procedure combining two moments of the actual data with a third moment of the logarithms of the observations. The recommended version of this method of mixed moments uses the mean and variance of the data as recorded and the mean of their logarithms. Equation (5.45) for $\mu_x$ and eqn (5.46) for $\sigma_x^2$ are therefore combined with eqn (5.35) for $\mu_z$. Rao (1980b) has presented a graphical approach for the solution of these three equations, but J. S. Colombi (personal communication, 1981) has shown that a direct solution is possible if $a$ is first calculated from the expression

$$\ln[\{(1-a)^2/(1-2a)\}^s(1-a)^{-1}]-a=0 \tag{5.48}$$

where

$$s=(\mu_z-\ln\mu_x)/\ln(1+\{\sigma_x^2/\mu_x^2\}) \tag{5.49}$$

and then $b$ and $c$ are computed from

$$b=(\mu_z-\ln\mu_x)/\{\ln(1-a)+a\} \tag{5.50}$$

$$c=\mu_z-ab \tag{5.51}$$

### 5.4.5. The General Extreme Value Distribution

The CDF of the general extreme value distribution, as expressed by Jenkinson (1955), may be written as

$$F(x)=\exp\{-[1-k(x-c)/a]^{1/k}\} \tag{5.52}$$

This equation may be used to describe three distinct types of variables. As $k$ tends to zero, the CDF of the extreme value type 1 (or Gumbel) distribution is obtained:

$$F(x_1)=\exp\{-\exp[-(x_1-c)/a]\} \tag{5.53}$$

for which the PDF is

$$f(x_1)=a^{-1}\exp\{-[(x_1-c)/a]-\exp[(x_1-c)/a]\} \tag{5.54}$$

The mean, $\mu_1$, and variance, $\sigma_1^2$, of this distribution are related to the location and scale parameters, $c$ and $a$, by the equations

$$\mu_1=c+0{\cdot}5772a \tag{5.55}$$

$$\sigma_1^2=\pi^2a^2/6 \tag{5.56}$$

and the skewness has a constant value of 1·14. The standardised variate, $y_1$, is related to $x_1$ by

$$y_1=(x_1-c)/a \tag{5.57}$$

and its CDF may be written as

$$G(y_1) = \exp\{-\exp(-y_1)\} \tag{5.58}$$

Since eqn (5.58) contains no parameters, $y_1$ and $G(y_1)$ may be related by means of a simple one-way table. In addition, since $G(y_1)$ is the probability of non-exceedance of the variate $y_1$, and may be expressed in terms of return period according to eqn (5.1), eqn (5.58) may be written as

$$y_1 = -\ln[-\ln\{1-(1/T)\}] \tag{5.59}$$

When the value of $k$ in eqn (5.52) is negative, the extreme value type 2 (or Fréchet) distribution is obtained. If the variable $z = \ln(x_2)$ follows a Gumbel distribution, then $x_2$ is a type 2 variable. The type 2 standardised variate is given by

$$y_2 = 1 - \{k(x_2 - c)/a\} \tag{5.60}$$

so that its CDF must be tabulated separately for different values of $k$.

When the $k$-value in eqn (5.52) is positive, the extreme value type 3 distribution is obtained. For the latter, $k>0$, $a>0$ and $-\infty \leqslant x_3 \leqslant c + a/k$. The standardised variate is given by

$$y_3 = \{k(x_3 - c)/a\} - 1 \tag{5.61}$$

which again has a CDF which must be tabulated separately for different values of $k$.

When the Gumbel distribution is applied, the value of $x_1$ corresponding to return period $T$ may be written as a function of $\mu_1$ and $\sigma_1$ instead of $c$ and $a$:

$$x_1(T) = \mu_1 + K(T)\sigma_1 \tag{5.62}$$

where the frequency factor, $K(T)$, is related to the standardised variate by the expression

$$K(T) = 0{\cdot}78y_1(T) - 0{\cdot}45 \tag{5.63}$$

Some authors (for example, Bruce and Clark, 1966) have tabulated $K(T)$ as a function of sample size. However, Lettenmaier and Burges (1982) have demonstrated that such factors lead to overestimation of the $T$-year flood and their use should therefore be avoided. As part of the same study, those authors have also shown that the method of maximum likelihood provided only marginal improvement over the method of moments in the variability of estimates obtained from the type 1 distribution.

A statistical test for discriminating between type 1 and type 2 distributions has been outlined by van Montfort (1970).

## 5.5. PARTIAL DURATION SERIES ANALYSIS

This approach involves the application of what are referred to as the 'peaks over a threshold' (POT) models. In its simplest form, this type of model requires the abstraction of the $M$ independent events within a record length of $N$ years ($M > N$), all of which exceed a threshold of discharge of magnitude $q_0$. A variety of such models is available, ranging from that in which the variation between years and between seasons in the number of values exceeding the threshold is ignored and a constant number of exceedances, $\xi = M/N$, is assumed to occur each year, to that in which the distribution of event magnitudes is allowed to vary from season to season. However, a feature common to all POT models is the assumption that the probability of exceedance of the $T$-year flood, $q(T)$, given that the threshold discharge is exceeded, i.e. the conditional probability, is exponential in form:

$$P\{q \geqslant q(T) \mid q(T) \geqslant q_0\} = \exp\left[-(q(T) - q_0)/a\right] \qquad (5.64)$$

where $P$ is the probability function and $a$ is the parameter value to be determined.

Perhaps the most widely used of the POT models is based upon the assumption that the number of values exceeding the threshold each year may be considered to be a random variable conforming to a Poisson distribution with parameter $\xi$ (see Shane and Lynn, 1964):

$$P_i = (\xi^i/i!) \exp(-\xi) \qquad (5.65)$$

where $P_i$ is the probability of having $i$ peaks over the threshold in a year. Using this assumption, the distribution of the number of values exceeding a higher level $q$ may be shown to be another Poisson variable with a parameter value equal to $\xi$ multiplied by the conditional probability given by eqn (5.64). Since Poisson variables are additive, the number of values exceeding $q$ during $\xi T$ years is a Poisson variable with a mean equal to $\xi T$ times the same conditional probability. When the latter quantity is unity, $q$ is the $T$-year flood, i.e.

$$P\{q \geqslant q(T) \mid q(T) \geqslant q_0\} = 1/(\xi T) \qquad (5.66)$$

Substituting this value in eqn (5.64), the $T$-year event is given by

$$q(T) = q_0 + a \ln \xi + a \ln T \qquad (5.67)$$

This relationship between flood magnitude and return period is also obtained if the number of exceedances per year is constant at $\xi$ and not a Poisson variable (see Cunnane, 1979).

There are two distinct methods of abstracting data for a POT analysis, and the estimation of parameters (and therefore the estimation of the $T$-year event) depends upon the abstraction procedure. In the first method the threshold $q_0$ is chosen beforehand, and the $M$ events are treated as a random sample from an exponential distribution with $q_0$ known but $a$ unknown. Both moment and maximum likelihood methods give $a$ as

$$\hat{a} = \bar{q} - q_0 \qquad (5.68)$$

where $\bar{q}$ is the arithmetic mean of the $M$ events. The $T$-year flood is computed from

$$q(T) = q_0 + \hat{a} \ln \hat{\xi} + \hat{a} \ln T \qquad (5.69)$$

where $\hat{\xi} = M/N$. In the second method the number of peaks, $M$, is predetermined, so that $\xi$ is known but $q_0$ and $a$ must be estimated from the sample. The maximum likelihood estimates of the latter, corrected for bias, are

$$\hat{a} = M(\bar{q} - q_{\min})/(M+1) \qquad (5.70)$$

$$\hat{q}_0 = q_{\min} - (\hat{a}/M) \qquad (5.71)$$

where $q_{\min}$ is the smallest of the $M$ peaks. The $T$-year flood is then estimated from

$$q(T) = \hat{q}_0 + \hat{a} \ln \xi + \hat{a} \ln T \qquad (5.72)$$

With both of these abstraction methods, the assumption that the underlying distribution of flood peaks is exponential should be checked graphically. Plotting positions for each event, $F_i$, may be allocated using the Gringorten formula (eqn (5.15) with $\alpha = 0{\cdot}44$), and standardised variates computed from

$$y_i = -\ln(1 - F_i) \qquad (5.73)$$

Since there are $\xi$ values exceeding $q_0$ each year and $\xi T$ values in $T$ years, the $T$-year event will occur once on average among every $\xi T$ values greater than $q_0$ and will have a return period $T' = \xi T$ sampling units, where a sampling unit is a peak over the threshold. The plotting positions corresponding to the values of $q(T)$ will therefore be associated with a return period of $T'$ sampling units, and their standardised variates will be given by $\ln(\xi T)$. Any pronounced curvature in the plot for flows approaching $q_0$ may be removed either by raising the value of $q_0$ (first method) or lowering the value of $M$ (second method).

## 5.6. CHOICE BETWEEN FREQUENCY DISTRIBUTIONS

Given a sample of annual floods, any of the distributions described briefly in Section 5.4 might be regarded as being a suitable choice to characterise the observations. A number of formal statistical tests are available for assessing the goodness-of-fit of a distribution to a sample of recorded floods. Perhaps the most widely known of these is the chi-square test, the details of which are to be found in the majority of statistical textbooks and manuals (see, for example, Crow *et al.*, 1960, pp. 85–7). Another commonly applied test involves the Kolmogorov–Smirnov index. In the latter, the sample is ranked in increasing order of magnitude such that $q_i$ is the $i$th smallest observation, and $q_1$ and $q_n$ are the smallest and largest sample values respectively. The empirical distribution function of the data is then defined by

$$\begin{aligned} S(q) &= 0; \quad q < q_i \\ &= i/n; \quad q_i < q < q_{i+1} \\ &= 1; \quad q \geqslant q_n \end{aligned} \tag{5.74}$$

Since $S(q)$ changes at each value of $q_i$, the difference between the theoretical distribution, $F(q)$, and the empirical distribution has two values:

$$\Delta_i^+ = (i/n) - F(q) \tag{5.75}$$

$$\Delta_i^- = F(q) - (i-1)/n \tag{5.76}$$

The Kolmogorov–Smirnov index, $D$, is given by the maximum of the values $\Delta_i^+$ and $\Delta_i^-$, $i = 1, 2, \ldots, n$. The sampling distribution of $D$ depends only on $n$ and not the form and parameters of the theoretical distribution, $F(q)$. For $n \geqslant 35$, the upper 10, 5 and 1% points of the distribution of $D$ are given by $1{\cdot}22/\sqrt{n}$, $1{\cdot}36/\sqrt{n}$ and $1{\cdot}63/\sqrt{n}$ respectively. If the observations were not drawn from the assumed distribution, then these values of $D$ will be exceeded more frequently than 10, 5 and 1%

of the time. When the computed value of $D$ exceeds one of these critical values, the distribution under test is rejected at that level of confidence.

Despite the widespread use of these and other tests, their ability to indicate a clear choice between two or more candidate distributions is open to doubt, particularly when the number of observations is small. An alternative approach, involving the estimation of economic losses associated with either underdesign or overdesign, suggested by Slack *et al.* (1975), has already been noted briefly in Section 5.1 but poses additional problems in specifying appropriate loss functions. However, the suitability of different distributions for flood frequency analysis may also be evaluated by comparing the sampling properties of the mean, standard deviation and skewness from different sizes of data sets drawn from such distributions with those of the statistics computed from similar lengths of record of observed annual floods. The results of such an exercise, based upon the historical data from 1351 gauging stations in the United States, have been reported in a series of papers by Matalas *et al.* (1975), Wallis *et al.* (1977) and Landwehr *et al.* (1978, 1980).

Using the results of an extensive random sampling experiment, Matalas *et al.* (1975) were able to show that estimates of skewness obtained from historical flood series were more variable than those of samples drawn from any of the well-known frequency distributions, a phenomenon that they referred to as the 'condition of separation'. This effect could not be attributed to either the smallness of the sample size or autocorrelation within the annual flood series, but Wallis *et al.* (1977) later demonstrated that systematic changes in skewness over time or the mixing of different values of skewness within a region could cause separation. According to Landwehr *et al.* (1978), a high kurtosis is a necessary but not a sufficient condition for separation. Moreover, the five-parameter Wakeby distribution can also account for separation because of its flexibility in reproducing a long, stretched left-hand tail.

These results have at least two important implications for flood frequency analysis. Firstly, as pointed out by Landwehr *et al.* (1980), moment estimates of kurtosis are subject to downward bias similar to that which affects the computation of the skewness coefficient. Fitting procedures which rely on both of these statistics are therefore most unlikely to yield reliable results. The importance of tail behaviour has been confirmed in separate studies by Shen *et al.* (1980) and Ochoa *et al.* (1980). Those authors drew a distinction between 'light-tailed' behaviour, as exemplified by distributional forms such as $x^a \exp(-bx)$ of the gamma family or $\exp(-bx^n)$ of the normal family for which all moments exist, and those typified by $F(x) = 1 - cx^{-k}$, such as the type 2 extreme value distribution, which possess no moments of order greater than or equal to $k$. The latter are also referred to as 'Paretian-tailed' distributions. Ochoa *et al.* (1980) applied the test presented by van Montfort (1970) to annual flood series from 407 gauging stations in the United States and found that a type 1 distribution was rejected in favour of a type 2 in over 70% of cases. In practice, the use of a light-tailed distribution when a Paretian-tailed distribution is more suitable would tend to underestimate markedly the frequency of extreme events.

Secondly, the temporal variation in skewness coefficients that was identified by Wallis *et al.* (1977) as a possible cause of separation also serves to cast doubt upon the use of contour maps to derive regional estimates of skewness as a means of avoiding the bias associated with its computation from recorded data. Matalas *et al.* (1975) had previously suggested that such maps should be conditioned on sample size. Landwehr *et al.* (1978) also pointed out that, if the skewness of the logarithmically transformed data is mapped, the contours in 'real' space are not uniquely defined because the transformation is also dependent upon the coefficients of variation and kurtosis as well as the underlying distribution function. Practical problems in constructing regional skewness maps have also been discussed by Jackson (1981).

In summary, such is the conflicting evidence in favour of each of the more wider applied frequency distributions and the different fitting procedures that the imposition of a choice, such as that of the log Pearson type 3 distribution as reported by Benson (1968), would appear to have much in its favour.

## REFERENCES

BENSON, M. A. (1968) Uniform flood-frequency estimating methods for Federal Agencies. *Wat. Resour. Res.*, **4,** 891–908.

BETHLAHMY, N. (1977) Flood analysis by SMEMAX transformation. *Proc. Am. Soc. Civ. Engrs., J. Hydraul. Div.*, **103**(HY1), 69–78.

BOBEE, B. (1975) The log Pearson type 3 distribution and its application in hydrology. *Wat. Resour. Res.*, **11,** 681–9.

BOBEE, B. and ROBITAILLE, R. (1975) Correction of bias in the estimation of the coefficient of skewness. *Wat. Resour. Res.*, **11,** 851–4.

Bobee, B. and Robitaille, R. (1977) The use of the Pearson type 3 and log Pearson type 3 distributions revisited. *Wat. Resour. Res.*, **13,** 427–43.

Box, G. E. P. and Cox, D. R. (1964) An analysis of transformations. *J. Roy. Stat. Soc., Ser. B*, **24,** 297–343.

Bruce, J. P. and Clark, R. H. (1966) *Introduction to hydrometeorology* (Pergamon Press, Oxford) 324 pp.

Buckett, J. and Oliver, F. R. (1977) Fitting the Pearson type 3 distribution in practice. *Wat. Resour. Res.*, **13,** 851–2.

Burges, S. J., Lettenmaier, D. P. and Bates, C. L. (1975) Properties of the three-parameter log-normal probability distribution. *Wat. Resour. Res.*, **11,** 229–35.

Chander, S., Spolia, S. K. and Kumar, A. (1978) Flood frequency analysis by power transformation. *Proc. Am. Soc. Civ. Engrs., J. Hydraul. Div.*, **104**(HY11), 1495–1504.

Charbeneau, R. J. (1978) Comparison of the two- and three-parameter log-normal distributions used in streamflow synthesis. *Wat. Resour. Res.*, **14,** 149–50.

Cochrane, N. J. (1967) An engineering calculation of risk in the provision for the passage of floods during the construction of dams. In *Proc. Int. Congr. on Large Dams, 9th Congr., Istanbul* (ICOLD, Paris) Vol. 5, pp. 325–41.

Condie, R. (1977) The log Pearson type 3 distribution: the $T$-year event and its asymptotic standard error by maximum likelihood theory. *Wat. Resour. Res.*, **13,** 987–91.

Crow, E. L., Davis, F. A. and Maxfield, M. W. (1960) *Statistics manual* (Dover Publications, New York) 288 pp.

Cunnane, C. (1978) Unbiased plotting positions—a review. *J. Hydrol.*, **37,** 205–22.

Cunnane, C. (1979) A note on the Poisson assumption in partial duration series models. *Wat. Resour. Res.*, **15,** 489–94.

Greenwood, J. A., Landwehr, J. M., Matalas, N. C. and Wallis, J. R. (1979) Probability weighted moments: definition and relation to parameters of several distributions expressible in inverse form. *Wat. Resour. Res.*, **15,** 1049–54.

Hinkley, D. (1977) On quick choice of power transformation, *Appl. Stat.*, **26,** 67–9.

Houghton, J. C. (1978a) Birth of a parent: the Wakeby distribution for modelling flood flows. *Wat. Resour. Res.*, **14,** 1105–9.

Houghton, J. C. (1978b) The incomplete means estimation procedure applied to flood frequency analysis. *Wat. Resour. Res.*, **14,** 1111–15.

Jackson, D. R. (1981) WRC standard flood frequency guidelines. *Proc. Am. Soc. Civ. Engrs., J. Wat. Resour. Plan. and Man. Div.*, **107**(WR1), 211–24.

Jenkinson, A. F. (1955) The frequency distribution of the annual maximum (or minimum) values of meteorological elements. *Quart. J. Roy. Met. Soc.*, **87,** 158–71.

Kirby, W. (1974) Algebraic boundedness of sample statistics. *Wat. Resour. Res.*, **10,** 220–2.

Landwehr, J. M., Matalas, N. C. and Wallis, J. R. (1978) Some comparisons of flood statistics in real and log space. *Wat. Resour. Res.*, **14,** 902–20.

LANDWEHR, J. M., MATALAS, N. C. and WALLIS, J. R. (1979a) Probability weighted moments compared with some traditional techniques in estimating Gumbel parameters and quantiles. *Wat. Resour. Res.*, **15,** 1055–64.

LANDWEHR, J. M., MATALAS, N. C. and WALLIS, J. R. (1979b) Estimation of parameters and quantiles of Wakeby distributions. 1. Known lower bounds. *Wat. Resour. Res.*, **15,** 1361–72.

LANDWEHR, J. M., MATALAS, N. C. and WALLIS, J. R. (1979c) Estimation of parameters and quantiles of Wakeby distributions. 2. Unknown lower bounds. *Wat. Resour. Res.*, **15,** 1373–9.

LANDWEHR, J. M., MATALAS, N. C. and WALLIS, J. R. (1980) Quantile estimation with some more or less floodlike distributions. *Wat. Resour. Res.*, **16,** 547–55.

LETTENMAIER, D. P. and BURGES, S. J. (1982) Gumbel's extreme value I distribution: a new look. *Proc. Am. Soc. Civ. Engrs., J. Hydraul. Div.*, **108**(HY4), 502–14.

LINDLEY, D. V. and MILLER, J. C. P. (1961) *Cambridge elementary statistical tables* (Cambridge Univ. Press) 35 pp.

MATALAS, N. C. and WALLIS, J. R. (1973) Eureka! It fits a Pearson type 3 distribution. *Wat. Resour. Res.*, **9,** 281–9.

MATALAS, N. C., SLACK, J. R. and WALLIS, J. R. (1975) Regional skew in search of a parent. *Wat. Resour. Res.*, **11,** 815–26.

MONTFORT, M. A. J. VAN (1970) On testing that the distribution of extremes is of type-I when type-II is the alternative. *J. Hydrol.*, **11,** 421–7.

NATURAL ENVIRONMENT RESEARCH COUNCIL (1975) *Flood Studies Report.* Vol. I: *Hydrological studies* (NERC, London) 550 pp.

NOZDRYN-PLOTNICKI, M. J. and WATT, W. E. (1979) Assessment of fitting techniques for the log Pearson type 3 distribution using Monte Carlo simulation. *Wat. Resour. Res.*, **15,** 714–18.

OCHOA, I. D., BRYSON, M. C. and SHEN, H.-W. (1980) On the occurrence and importance of Paretian-tailed distributions in hydrology. *J. Hydrol.*, **48,** 53–62.

RAO, D. V. (1980a) Log Pearson type 3 distribution: a generalised evaluation. *Proc. Am. Soc. Civ. Engrs., J. Hydraul. Div.*, **106**(HY5), 853–72.

RAO, D. V. (1980b) Log Pearson type 3 distribution: method of mixed moments. *Proc. Am. Soc. Civ. Engrs., J. Hydraul. Div.*, **106**(HY6), 999–1019.

SANGAL, B. P. and BISWAS, A. K. (1970) The three-parameter lognormal distribution and its applications in hydrology. *Wat. Resour. Res.*, **6,** 505–15.

SHANE, R. M. and LYNN, W. R. (1964) Mathematical model for flood risk evaluation. *Proc. Am. Soc. Civ. Engrs., J. Hydraul. Div.*, **90**(HY6), 1–20.

SHEN, H.-W., BRYSON, M. C. and OCHOA, I. D. (1980) Effect of tail behaviour assumptions on flood prediction. *Wat. Resour. Res.*, **16,** 361–4.

SINGH, K. P. (1968) Hydrologic distributions resulting from mixed populations and their computer simulation. Int. Assoc. Sci. Hydrol., Publ. No. 81, pp. 671–81.

SLACK, J. R., WALLIS, J. R. and MATALAS, N. C. (1975) On the value of information in flood frequency analysis. *Wat. Resour. Res.*, **11,** 629–47.

STEDINGER, J. R. (1980) Fitting log-normal distributions to hydrologic data. *Wat. Resour. Res.*, **16,** 481–90.

TUNG, Y.-K. and MAYS, L. W. (1981a) Risk models for flood levee design. *Wat. Resour. Res.*, **17,** 833–41.

TUNG, Y.-K. and MAYS, L. W. (1981b) Optimal risk-based design of flood levee systems. *Wat. Resour. Res.*, **17,** 843–52.

TUNG, Y.-K. and MAYS, L. W. (1982) Optimal risk-based hydraulic design of bridges. *Proc. Am. Soc. Civ. Engrs., J. Wat. Resour. Plan. and Man. Div.*, **108**(WR2), 191–203.

VENUGOPAL, K. (1980) Flood analysis by SMEMAX transformation—a review. *Proc. Am. Soc. Civ. Engrs., J. Hydraul. Div.*, **106**(HY2), 338–40.

WALLIS, J. R., MATALAS, N. C. and SLACK, J. R. (1974) Just a moment! *Wat. Resour. Res.*, **10,** 211–19.

WALLIS, J. R., MATALAS, N. C. and SLACK, J. R. (1976) Effect of sequence length *n* on the choice of assumed distribution of floods. *Wat. Resour. Res.*, **12,** 457–71.

WALLIS, J. R., MATALAS, N. C. and SLACK, J. R. (1977) Apparent regional skew. *Wat. Resour. Res.*, **13,** 159–82.

YEN, B. C. (1970) Risks in hydrologic design of engineering projects. *Proc. Am. Soc. Civ. Engrs., J. Hydraul. Div.*, **96**(HY4), 959–66.

YEN, B. C., TANG, W. H. and MAYS, L. W. (1974) Designing storm sewers using the rational method. *Wat. and Sewage Wks.*, **121**(10), 92–5; ibid. **121**(11), 84–5.

# 6

# Rainfall–Runoff Relationships

## 6.1. INTRODUCTION

Following the more widespread use of digital computers in hydrology during the early 1960s, Amorocho and Hart (1964) commented upon the growth of two distinct approaches to the basic problem of establishing the relationship between rainfall and runoff, which they referred to as 'physical hydrology' and 'systems investigation'. The former term was used to describe investigations into the behaviour and interdependence of the component processes within the hydrological cycle, as described in Chapter 1. Within this category, individual studies may cover the working of only one component under a restricted range of conditions, i.e. the movement of water through a soil column, but their long-term objective is a complete synthesis of the hydrological cycle. Such investigations are not generally made with the object of providing data for the solution of engineering problems; interest is centred on the achievement of scientific understanding. In contrast, Amorocho and Hart regarded the primary concern of systems investigation as being the solution of technological problems within limited time periods, in which historical records of only those hydrological variables that appear significant to the overall behaviour of a catchment are employed. Relationships between those variables are developed to provide a solution which is at least applicable within the working range of the data. Although interest tends to be centred upon methods of solution, reference to the results of studies in physical hydrology is essential to both the selection of the pertinent variables and the interpretation of results.

Dooge (1968) has defined a system as any structure, device, scheme or procedure, real or abstract, that interrelates in a given time reference an input, cause or stimulus of matter, energy or information and an output, effect or response of information, energy or matter. The extent to which this concept is useful in a hydrological context may be appreciated by considering again the classical schematic representation of the hydrological cycle, as depicted in Fig. 1.1. As noted in Chapter 1, such diagrams tend to be misleading in that they do not draw sufficient attention to the interrelationships between the principal storages. Furthermore, the existence of sub-components to processes such as river flow is less than obvious. When Fig. 1.1 is redrawn in systems notation, as shown in Fig. 1.2, the component processes are more readily identified, and those portions of the hydrological cycle of interest to different professional groups more clearly delineated. For example, the section of the cycle concerned with the relationship between rainfall and runoff of direct interest to the engineering hydrologist is depicted in Fig. 1.2 by the broken lines.

The representation of the catchment system in terms of inputs and outputs is shown in Fig. 6.1. The inputs, $x_i(t)$, $i = 1, 2, \ldots, n$, are the records of the $n$ autographic raingauges within the drainage area, which produce an output of streamflow, $y(t)$, at the catchment outlet. The $x_i(t)$ are frequently averaged to produce a single sequence, $x(t)$, which is then referred to as a 'lumped' input. Since the relationship between $x(t)$ and $y(t)$ is then evaluated without reference to the physics of the component processes of the hydrological cycle which transform rainfall into runoff, the catchment is effectively treated as a 'black box', as illustrated in Fig. 6.1.

The techniques which have been employed to investigate the catchment system are summarised diagrammatically in Fig. 6.2. A major distinction may be drawn between linear and non-linear approaches. A

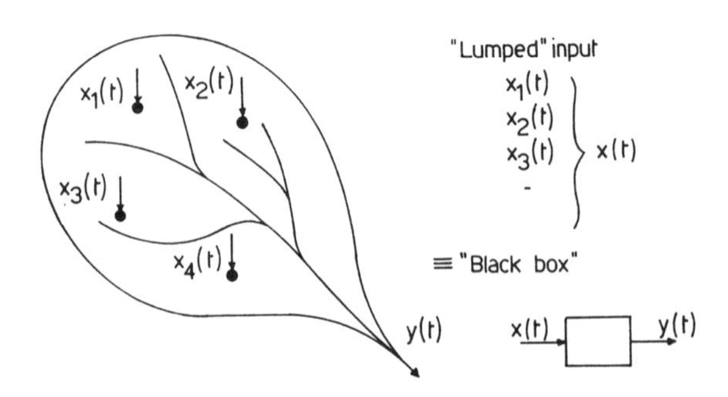


FIG. 6.1. The catchment system.

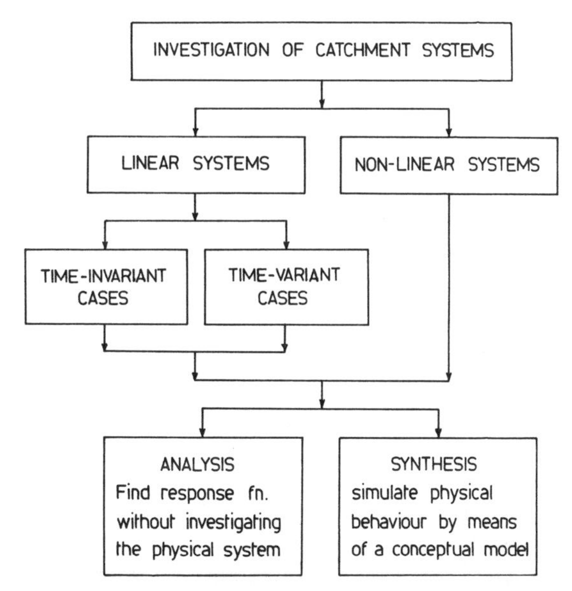


FIG. 6.2. The investigation of catchment systems.

linear system has the property that if an input $x_1$ produces an output $y_1$ and an input $x_2$ produces an output $y_2$, then an input $ax_1+bx_2$ produces an output $ay_1+by_2$ for all inputs $x_1$ and $x_2$ and pairs of constants, $a$ and $b$. Such a system is therefore governed by a linear equation, differential or otherwise. Since individual components of the hydrological cycle tend to exhibit some degree of non-linearity, the linear approach is inevitably an approximation in which fidelity of description is sacrificed for mathematical tractability.

Catchment systems may be further subdivided into time-invariant and time-variant cases. If an input $x(t)$ to a time-invariant linear system produces an output $y(t)$, then an input $x(t+T)$ produces an output $y(t+T)$ for all values of the time shift, $T$. The mathematical function which transforms the input into an output does not therefore change with time for such a system.

The methods which have been employed to identify and describe the response function of a catchment system may be considered to fall into two basic categories, referred to as analysis and synthesis. In the former, the response function is isolated without any attempt to investigate the catchment. In the latter, the available knowledge of the

**Table 6.1. Problems amenable to a systems approach**
(modified from Dooge, 1973)

| *Type of problem* | | *Input* | *Response function* | *Output* |
|---|---|---|---|---|
| ANALYSIS | Forecasting | Given | Given | Required |
| | Identification | Given | Required | Given |
| | Detection | Required | Given | Given |
| SYNTHESIS | | Given | Required | Given |

physical behaviour of the drainage area is used to construct a 'conceptual model' whose response function approximates that of the catchment.

The practical problems that can be handled conveniently using the systems approach are summarised in Table 6.1. Using the analysis approach, for example, the estimation of an output of streamflow from a catchment, given its response function and the rainfall input, may be termed forecasting. However, the more common problem encountered is that of identification, i.e. determination of the catchment response function from a knowledge of the rainfall input and streamflow output. Detection, the estimation of the rainfall input from the streamflow output and the catchment response function, is rarely required. With the synthesis approach, given rainfall input and streamflow output, a suitable conceptual model must be both identified and then fitted using the available data. If the model is non-linear in character, the difficulties of the latter operation are compounded by parameter estimation problems, as discussed below in Section 6.4.

One of the principal tools for the forecasting of streamflows which has now been in widespread use for over 50 years is the Unit Hydrograph Method as introduced by Sherman (1932). The basis of the Unit Hydrograph Method has already been outlined in Chapter 4. In effect, the method is based upon a simplified catchment model, of the form illustrated in Fig. 6.3, in which the incident precipitation, $P$, is routed down 'fast' and 'slow' response paths. The portion taking the former route becomes the effective rainfall or rainfall excess, $P_e$, which is converted into surface runoff, $Q_s$, by the direct storm response in the form of the unit hydrograph. The residual, $P-P_e$, becomes the infiltration, $F$, to the upper soil horizons. The division between $P_e$ and $F$ is largely governed by the level of soil moisture storage, giving rise to an

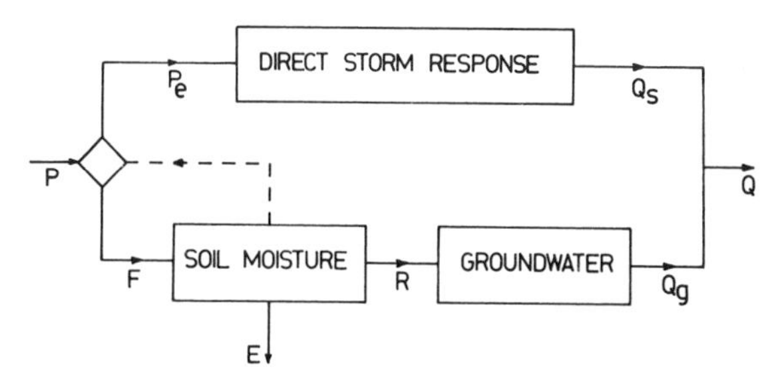


FIG. 6.3. Simplified catchment model (modified from Dooge, 1973).

important feedback mechanism. The simplified model ignores the possibility of interflow or throughflow occurring, so that the soil moisture storage is depleted either by evaporation, $E$, or percolation to groundwater, $R$. The latter gives rise to the baseflow, $Q_g$. Provided that the effective rainfall can be separated from the total rainfall, $P$, and the surface runoff from the total discharge, $Q$, attention may be confined to the direct storm response. The determination of the latter using the analysis and synthesis approaches is considered in Sections 6.2 and 6.3 respectively.

## 6.2. ANALYSIS APPROACH

One of the most common methods of analysis applied to the identification problem is that of transformation, the basis of which is illustrated in Fig. 6.4. This approach exploits the possibility that, in the transformed domain, a simple analytical relationship between the input $x(t)$, the output $y(t)$ and the system response $h(t)$ is available that can be inverted to obtain $h(t)$. A ready example of transformation is provided by the use of orthogonal functions.

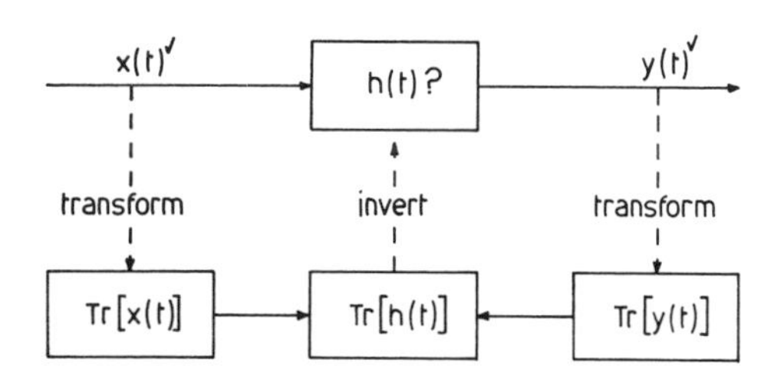


FIG. 6.4. The transformation approach to the identification problem.

In general, any function $g(t)$ can be represented exactly by the sum of an infinite series of other functions, $f(t)$, e.g.

$$g(t) = c_0 f_0(t) + c_1 f_1(t) + c_2 f_2(t) + \cdots \tag{6.1}$$

where the $c_m$, $m = 0, 1, 2, \ldots$, are coefficients. Orthogonal functions have the particular property that

$$\begin{aligned} \int_a^b f_m(t) f_n(t)\, \mathrm{d}t &= K \quad \text{if} \quad m = n \\ &= 0 \quad \text{if} \quad m \neq n \end{aligned} \tag{6.2}$$

where $a$, $b$, $K$ depend upon the particular $f(t)$. The coefficients $c_m$ of an orthogonal series are therefore easily found from

$$\int_a^b g(t) f_m(t)\, \mathrm{d}t = 0 + 0 \cdots + c_m K + 0 + \cdots$$

$$c_m = \frac{1}{K} \int_a^b g(t) f_m(t)\, \mathrm{d}t \tag{6.3}$$

In applying the transformation approach to the identification of the response function of a catchment, the rainfall input, $x(t)$, the streamflow output, $y(t)$, and the system response, $h(t)$, are all expressed in terms of orthogonal series:

$$x(t) = \sum_{m=0}^{\infty} a_m f_m(t) \tag{6.4}$$

$$y(t) = \sum_{m=0}^{\infty} A_m f_m(t) \tag{6.5}$$

$$h(t) = \sum_{m=0}^{\infty} \alpha_m f_m(t) \tag{6.6}$$

The convolution integral, given by eqn (4.6), which expresses the output, $y(t)$, in terms of the input, $x(t)$, and the system response, $h(t)$, then provides the basis for deriving general linkage equations between the coefficients, $\alpha_m$, and the $a_m$ and $A_m$. Given the $x(t)$ and $y(t)$ for a storm event, the $a_m$ and $A_m$ may be computed according to eqn (6.3). The $\alpha_m$ are then obtained from the $a_m$ and $A_m$ using the general linkage equations, and the $h(t)$ reconstituted from eqn (6.6).

Among the first applications of the transformation approach was that by O'Donnell (1960), who proposed the use of Fourier series. Since the output, $y(t)$, has a longer time base than the input, $x(t)$, in hydrological applications, the 'period', $L$, of the functions is equated to the former:

$$y(t)=A_0+\sum_{r=1}^{\infty}\left[A_r\cos r\frac{2\pi t}{L}+B_r\sin r\frac{2\pi t}{L}\right] \tag{6.7}$$

The input and systems response are similarly expressed:

$$x(t)=a_0+\sum_{r=1}^{\infty}\left[a_r\cos r\frac{2\pi t}{L}+b_r\sin r\frac{2\pi t}{L}\right] \tag{6.8}$$

$$h(t)=\alpha_0+\sum_{r=1}^{\infty}\left[\alpha_r\cos r\frac{2\pi t}{L}+\beta_r\sin r\frac{2\pi t}{L}\right] \tag{6.9}$$

The coefficients, $A_r$ and $B_r$, are computed according to the standard relationships for orthogonal functions:

$$A_r=\frac{2}{L}\int_0^L y(t)\cos r\frac{2\pi t}{L}\,\mathrm{d}t;\quad \text{but}\quad A_0=\frac{1}{L}\int_0^L y(t)\,\mathrm{d}t$$

$$B_r=\frac{2}{L}\int_0^L y(t)\sin r\frac{2\pi t}{L}\,\mathrm{d}t \tag{6.10}$$

and similar expressions are applicable for the $a_r$ and $b_r$. O'Donnell (1960) showed that the coefficients, $\alpha_r$ and $\beta_r$, defining the instantaneous unit hydrograph (IUH) of the catchment, $h(t)$, could be computed from the $A_r$, $B_r$, $a_r$ and $b_r$ using the linkage equations

$$\alpha_r=\frac{2}{L}\frac{a_rA_r+b_rB_r}{a_r^2+b_r^2};\quad \text{but}\quad \alpha_0=\frac{1}{L}\frac{A_0}{a_0}$$

$$\beta_r=\frac{2}{L}\frac{a_rB_r-b_rA_r}{a_r^2+b_r^2} \tag{6.11}$$

When applying this technique in practice, the following restrictions apply:

(1) the input and output can only be sampled at a finite number of time points (even a continuous chart can only be read in such a fashion); and
(2) the input data are normally expressed as total volumes within successive equal time increments, i.e. as a histogram and not a continuous intensity function.

The integrals of eqn (6.10) cannot be performed analytically with sampled data; nor can an infinite series of equations be established. The integrals must therefore be replaced by summations and the infinite Fourier series by finite harmonic series. Since the series are orthogonal under summation as well as integration, eqns (6.11) also apply to the latter. Any $n$ data points may be fitted exactly by an $n$-term harmonic series, although such a series cannot be expected to fit the function exactly between the data points. This approach yields the $n$ ordinates of the finite-period unit hydrograph (TUH) rather than the IUH (see O'Donnell, 1966). A set of expressions similar in form to eqns (6.11) has been derived by Diskin (1977) for Laguerre functions.

Another commonly used method of identification is based upon the technique of matrix inversion. Apparently first suggested by Snyder (1955), the method was expressed in computer-compatible form by the Tennessee Valley Authority (1961). Given the total volumes of rainfall within successive time increments of $T$ hours, $x_i$, $i=1,2,\ldots,m$, and the ordinates of the TUH at the same discrete time points, $h_j$, $j=1,2,\ldots,n$, the ordinates of the surface runoff may be computed from the following $p=m+n-1$ equations:

$$\begin{aligned}
y_1 &= x_1h_1 \\
y_2 &= x_2h_1 + x_1h_2 \\
y_3 &= x_3h_1 + x_2h_2 + x_1h_3 \\
&\cdots \\
y_m &= x_mh_1 + x_{m-1}h_2 + \cdots + x_1h_m \\
y_{m+1} &= x_mh_2 + \cdots + x_2h_m + x_1h_{m-1} \\
&\cdots \\
y_{m+n-2} &= x_mh_{n-1} + x_{m-1}h_n \\
y_{m+n-1} &= x_mh_n
\end{aligned} \tag{6.12}$$

When recast in matrix form, these $m+n-1$ equations for the $n$ unknown TUH ordinates may be summarised in the equation

$$\mathbf{XH} = \mathbf{Y} \tag{6.13}$$

where $\mathbf{X}$ is a rectangular matrix with $n$ columns and $p=m+n-1$ rows:

$$\mathbf{X} = \begin{bmatrix}
x_1 & 0 & 0 & & \cdots & 0 & \cdots & 0 \\
x_2 & x_1 & 0 & & \cdots & 0 & \cdots & 0 \\
x_3 & x_2 & x_1 & & \cdots & 0 & \cdots & 0 \\
\cdots & & & & & & & \cdots \\
x_m & x_{m-1} & x_{m-2} & \cdots & x_1 & 0 & \cdots & 0 \\
0 & x_m & x_{m-1} & \cdots & x_2 & x_1 & \cdots & 0 \\
\cdots & & & & & & & \cdots \\
0 & 0 & 0 & & \cdots & 0 & x_m & x_{m-1} \\
0 & 0 & 0 & & \cdots & 0 & 0 & x_m
\end{bmatrix} \tag{6.14}$$

**H** is a column vector with $n$ rows such that

$$\mathbf{H} = [h_1\ h_2 \cdots h_n]^{\mathrm{T}} \tag{6.15}$$

where T denotes the operation of matrix transposition; and **Y** is a column vector with $p$ rows such that

$$\mathbf{Y} = [y_1\ y_2 \cdots y_m\ y_{m+1} \cdots y_{m+n-1}]^{\mathrm{T}} \tag{6.16}$$

In order to invert eqn (6.13), **X** is multiplied by its transpose, $\mathbf{X}^{\mathrm{T}}$, which is obtained by interchanging the rows and columns of **X**, to give the square matrix $\mathbf{W} = \mathbf{X}^{\mathrm{T}}\mathbf{X}$. Then

$$\mathbf{X}^{\mathrm{T}}\mathbf{X}\mathbf{H} = \mathbf{W}\mathbf{H} = \mathbf{X}^{\mathrm{T}}\mathbf{Y}$$

and

$$\mathbf{H} = \mathbf{W}^{-1}\mathbf{X}^{\mathrm{T}}\mathbf{Y} \tag{6.17}$$

where the $-1$ indicates the operation of matrix inversion. This matrix manipulation automatically provides the least-squares solution for the $n$ TUH ordinates from the $p$ equations, as the following development will demonstrate. Let **R** be a column vector with $p$ rows such that

$$\mathbf{R} = \mathbf{X}\mathbf{H} - \mathbf{Y} \tag{6.18}$$

and let

$$S = \sum_{i=1}^{p} r_i^2 \tag{6.19}$$

where $r_i$ is the $i$th element of **R**. The least-squares solution is that for which $\partial S/\partial h_j = 0$, $j = 1, 2, \ldots, n$, where $h_j$ is the $j$th element of the vector **H**. Now

$$\frac{\partial S}{\partial h_j} = 2\sum_{i=1}^{p} r_i \frac{\partial r_i}{\partial h_j}; \qquad j = 1, 2, \ldots, n \tag{6.20}$$

but

$$r_i = \sum_{j=1}^{n} (x_{ij}h_j) - y_i; \qquad \frac{\partial r_i}{\partial h_j} = x_{ij} \tag{6.21}$$

where $x_{ij}$ is the $j$th element in the $i$th row of the matrix **X**. Therefore

$$\begin{aligned} &r_1x_{11} + r_2x_{21} + r_3x_{31} + \cdots + r_px_{p1} = 0 \\ &\cdots \\ &r_1x_{1n} + r_2x_{2n} + r_3x_{3n} + \cdots + r_px_{pn} = 0 \end{aligned} \tag{6.22}$$

or, in matrix notation,

$$\mathbf{Z}\mathbf{R} = 0 \tag{6.23}$$

where $x_{ij}$ is the $i$th element in the $j$th row of $\mathbf{Z}$, i.e.

$$\mathbf{Z} = \mathbf{X}^{\mathrm{T}} \tag{6.24}$$

so that

$$\mathbf{X}^{\mathrm{T}}\mathbf{R} = \mathbf{X}^{\mathrm{T}}(\mathbf{XH} - \mathbf{Y}) = 0 \tag{6.25}$$

from eqn (6.18). Solving eqn (6.25) for $\mathbf{H}$ gives

$$\mathbf{H} = (\mathbf{X}^{\mathrm{T}}\mathbf{X})^{-1}\mathbf{X}^{\mathrm{T}}\mathbf{Y} \tag{6.26}$$

which is identical to eqn (6.17).

Another potentially useful method of identification, which appears to have received comparatively little attention, is that proposed by De Laine (1970). This method is applicable to cases in which both the system response ordinates, $h_j$, $j = 1, 2, \ldots, n$, and the rainfall inputs, $x_i$, $i = 1, 2, \ldots, m$, are unknown. If input, output and system response are all standardised, i.e. each ordinate is expressed as a proportion of the sum of all the ordinates, such that

$$\sum_{i=1}^{m} x_i = \sum_{j=1}^{n} h_j = \sum_{k=1}^{p} y_k = 1 \tag{6.27}$$

then eqns (6.12) and (6.27) provide a system of $m + n$ equations with $m + n$ unknowns. Since rainfall and runoff observations are generated in a real physical situation, there must be at least one set of real values which satisfies these equations, but their determination is hampered by the existence of an unknown number of real and complex solutions. This problem may be avoided by considering a second output and obtaining a second group of solutions. In each of the two groups there must be a common set of the $h_j$, since the outputs were produced by the same system. Given the $h_j$, the inputs can be found from the outputs by substitution.

The method of solution proposed by De Laine (1970) involves expressing the successive ordinates of the input and system response as the coefficients of a polynomial in a dummy variable, $s$ (say):

$$x_s = x_1 + x_2 s + x_3 s^2 + \cdots + x_m s^{m-1} \tag{6.28}$$

$$h_s = h_1 + h_2 s + h_3 s^2 + \cdots + h_n s^{n-1} \tag{6.29}$$

Multiplying eqn (6.28) by eqn (6.29) then yields

$$x_s h_s = x_1 h_1 + (x_1 h_2 + x_2 h_1)s + (x_1 h_3 + x_2 h_2 + x_3 h_1)s^2 + \cdots + x_m h_n s^{m+n-2} \tag{6.30}$$

Reference to eqns (6.12) shows that eqn (6.30) reduces to

$$x_s h_s = y_1 + y_2 s + y_3 s^2 + \cdots + y_{m+n-1} s^{m+n-2} \tag{6.31}$$

The polynomial whose coefficients are the successive ordinates of the output is therefore the product of two polynomials whose coefficients are the successive ordinates of the input and system response respectively. Therefore, the factors of the polynomial in $y$ are also the factors of the polynomials in $x$ and $h$. However, these factors cannot be allocated to eqns (6.28) and (6.29) until the polynomial corresponding to a second set of output data is factorised. The factors of the polynomial $h_s$ will then be common to both sets of factors derived from the two polynomials in $y$. The method therefore reduces to finding the common roots from two (or more) polynomials in $y$ equated to zero, and the reconstitution of the polynomial $h_s$ from those roots. The coefficients of the polynomial then provide estimates of the successive ordinates of the system response. Both operations are greatly assisted by the availability of standard computer software both to obtain the roots of a polynomial function and to reconstitute a polynomial from a set of factors. The difficulty of the approach lies in the initial selection of common roots. Storms having similar shapes of runoff hydrograph (and therefore similar net rainfall patterns) should be avoided, since all roots of the polynomials $y$ could be paired off. The use of at least three dissimilar shapes of hydrograph mitigates this difficulty, but scope remains for the derivation of some objective method for pairing roots.

Although De Laine (1970) presented the above approach as the means of obtaining the systems response without using rainfall data, the method could be applied by factoring the polynomials in $x$ and $y$, and eliminating the former from the latter. This alternative technique has yet to be fully explored.

## 6.3. SYNTHESIS APPROACH

In contrast to the analysis approach, synthesis involves the prior assumption of a conceptual model, the response function of which is matched to that of the catchment area. Such conceptual models may vary greatly in complexity, but one of the simplest is the 'single linear reservoir', which is characterised by a storage–outflow relationship of the form

$$S = KQ \tag{6.32}$$

where $S$ is the storage corresponding to an outflow discharge of $Q$, and $K$ is a constant having units of time. When eqn (6.32) is combined with the continuity equation

$$P - Q = \mathrm{d}S/\mathrm{d}t \tag{6.33}$$

a first-order differential equation is obtained which, assuming $Q = 0$ when time $t = 0$, yields the solution

$$Q = P[1 - \exp(-t/k)] \tag{6.34}$$

Furthermore, assuming that the rainfall, $P$, ceases when $t = T$ and $Q = Q_0$, the following expression is obtained for the recession limb of the hydrograph:

$$Q = Q_0 \exp[-(t - T)/K] \tag{6.35}$$

If the input $P$ satisfies the storage $S$ instantaneously making $T = 0$, $Q_0 = S/K$ and, for a unit input, the form of the instantaneous unit hydrograph (IUH) is obtained as

$$h(0, t) = (1/K) \exp(-t/K) \tag{6.36}$$

The significance of the parameter $K$ in eqn (6.32) may be evaluated by taking moments about the time origin for the input and output and noting that $\mathrm{d}S/\mathrm{d}t = K\,\mathrm{d}Q/\mathrm{d}t$:

$$\int_0^\infty Pt\,\mathrm{d}t - \int_0^\infty Qt\,\mathrm{d}t = \int_0^\infty Kt\frac{\mathrm{d}Q}{\mathrm{d}t}\,\mathrm{d}t \tag{6.37}$$

Evaluating the right-hand side of eqn (6.37) and applying the condition $Q = 0$ at $t = \infty$, the following expression is obtained for $K$:

$$K = \left[\int_0^\infty Qt\,\mathrm{d}t \Big/ \int_0^\infty Q\,\mathrm{d}t\right] - \left[\int_0^\infty Pt\,\mathrm{d}t \Big/ \int_0^\infty Q\,\mathrm{d}t\right] \tag{6.38}$$

However, by continuity,

$$\int_0^\infty Q\,\mathrm{d}t = \int_0^\infty P\,\mathrm{d}t \tag{6.39}$$

so that according to eqn (6.38), the constant, $K$, is equal to the time difference between the centroids of the effective rainfall distribution and the hydrograph of surface runoff.

The single linear reservoir has proved to be an effective simple conceptual model for many applications, particularly in the investigation of urban catchments. However, its utility is limited by its lack of flexibility in having only one parameter, $K$, to vary when fitting the model to recorded data. The introduction of a second parameter, using the single linear reservoir as a building block, considerably extends the scope of this model. For example, Nash (1957, 1960) introduced the 'cascade of linear reservoirs' in which the output from the first reservoir becomes the input for the second, and the output from the second becomes the input to the third, and so on through the cascade without further inputs to intermediate reservoirs. Assuming an instantaneous unit input to the first reservoir, its output may be obtained from eqn (6.36), i.e.

$$Q_1 = (1/K) \exp(-t/K) \tag{6.40}$$

Using $Q_1$ as the input to the second reservoir, its output, $Q_2$, may be found to be

$$Q_2 = (t/K^2) \exp(-t/K) \tag{6.41}$$

Repeated routing leads to the general result that the output, $Q_n$, from the $n$th reservoir, which is also the IUH for $n$ reservoirs each having a constant, $K$, is given by

$$h(0, t) = [1/K\Gamma(n)](t/K)^{n-1} \exp(-t/K) \tag{6.42}$$

where $\Gamma$ denotes the gamma function.

The two parameters, $n$ and $K$, may be expressed in terms of the moments about the time origin of the effective rainfall distribution and the surface runoff hydrograph. The $m$th moment about the origin of $h(0, t)$ is given by

$$M_m = [1/K\Gamma(n)] \int_0^\infty (t/K)^{n-1} \exp(-t/K) t^m \, dt \tag{6.43}$$

After some algebra, eqn (6.43) reduces to

$$M_m = K^m [\Gamma(m+n)/\Gamma(n)] \tag{6.44}$$

For the first moment, $M_1$ is simply equal to the product $nK$. Therefore

$$M_{1Q} - M_{1P} = nK \tag{6.45}$$

where $M_{kQ}$, $M_{kP}$ are the $k$th moments about the time origin of the surface runoff hydrograph and the effective rainfall distribution respectively. For $m = 2$, equating the moments yields

$$M_{2Q} - M_{2P} - 2nKM_{1P} = n(n-1)K^2 \tag{6.46}$$

Equations (6.45) and (6.46) provide two equations with two unknowns, $n$ and $K$, which form the basis for model fitting in what is effectively another transform method, with $P$ and $Q$ being replaced by their moments.

The application of alternative two-parameter conceptual models, such as the cascade of two unequal linear reservoirs (Sarginson and Bourne, 1969), and the alternate parameter linear reservoir (Swinnerton *et al.*, 1973) are considered further in Chapters 8 and 9. Another two-parameter conceptual model may be constructed by combining the single linear reservoir with another device known as a 'linear channel'. In the latter, an input at time $t$, $Q(t)$, becomes an output, $Q(t+T)$ at time $t+T$, i.e. the hydrograph is delayed by $T$ without change in shape. The linear channel plus linear reservoir model has been investigated by Viessman *et al.* (1970).

The above models provide typical examples of what may be termed the linear synthesis approach. If the structure of the conceptual model is complicated by the introduction of threshold values on the storages, or the storage–discharge relationships of the reservoirs within the model do not follow eqn (6.32), the approach becomes non-linear. A typical example of a non-linear conceptual model is provided by the O'Donnell model (Dawdy and O'Donnell, 1965), a representation of which is provided in Fig. 6.5. This model was devised primarily to test automated methods of parameter determination (see Section 6.4 below) and not as an operational model. Nevertheless, the structure exhibits all the characteristics that would be expected in such a model.

As shown in Fig. 6.5, the O'Donnell model has four storage elements, consisting of:

(1) a surface storage, $R$, augmented by rainfall, $P$, and depleted by evaporation, $E$, infiltration, $F$, and (when $R$ exceeds the threshold, $R_0$) overland flow, $Q_r$;
(2) a channel storage, $S$, augmented by $Q_r$ and depleted by storm runoff, $Q_s$;
(3) a soil moisture storage, $M$, augmented by $F$ and capillary rise, $C$, and depleted by evapotranspiration, $T$, and (when $M$ exceeds the threshold, $M_0$) percolation, $D$; and
(4) a groundwater storage, $G$, augmented by $D$, and depleted by $C$ and baseflow, $Q_g$; when $G$ exceeds the threshold $G_0$, $M$ is absorbed into $G$, $C$ and $D$ no longer operate and $T$ and $F$ act only on $G$.

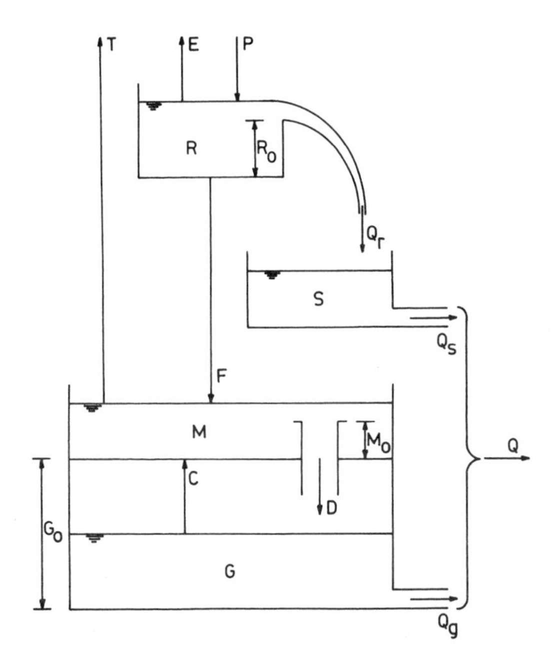


FIG. 6.5. The O'Donnell model.

This model has nine parameters: $R_0$; three parameters in the equation for infiltration; the constant of the linear reservoir, $S$; $M_0$ and $G_0$; the upper bound for $C$; and the constant of the linear reservoir, $G$. When calibrating this type of model, the use of records beginning with a long, dry period is recommended so that the four storage elements can be allocated zero values and the potential infiltration rate set to its maximum. The nine parameter values must then be determined from records of rainfall, streamflow and potential evaporation, using either trial-and-error methods or automatic optimisation procedures. The former methods demand a subtle combination of experience and intuition on the part of the operator, since obviously the temporal variations of the output streamflow are more sensitive to some parameters than to others. In contrast, optimisation or hill-climbing techniques involve the application of a computer algorithm which determines the values of the model parameters that maximise (or minimise) some error criterion or objective function.

Whether manual or machine methods are employed, and however

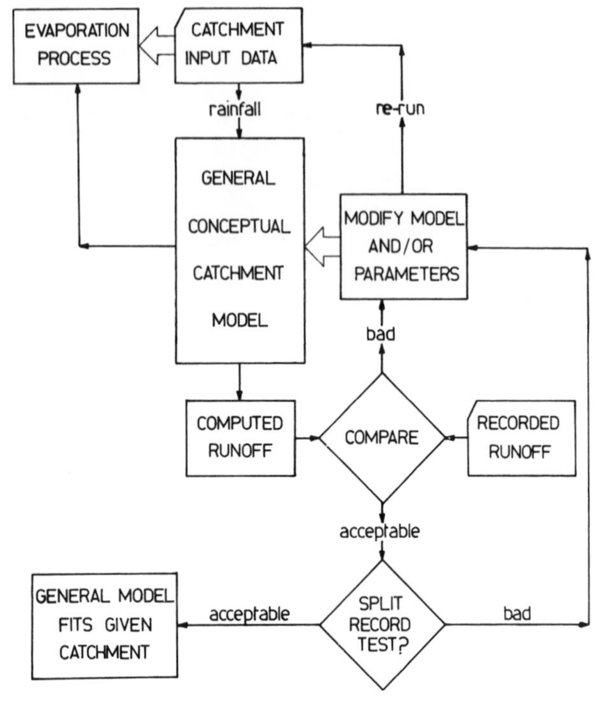


FIG. 6.6. Procedure for system synthesis.

simple or complex the conceptual model, the systems synthesis approach is the same (see Fig. 6.6). The conceptual model is applied to the input rainfall sequence to produce time series of evaporation and streamflow. The former is compared with the potential evaporation record for the catchment and adjustments made to the actual evaporation, if necessary. The synthesised records of streamflow are compared with the observations, and the model parameters modified if the objective function is not satisfied. When parameter values have been found that provide a satisfactory fit of the model to the data, a second phase of testing is entered in which the model is used to synthesise streamflows from storm events that were not employed in the calibration exercise. This 'split-record testing' provides a true, objective test of model performance whereas the reproduction of events included in the calibration merely confirms the validity of the arithmetic.

In system synthesis, the selection of the objective function is equally if not more important than the choice of model structure. Diskin and Simon (1977) have stressed that the two cannot be considered in isolation. Given a time series of observed flows $q_i$, $i = 1, 2, \ldots, m$, and a model output $\hat{q}_i$, $i = 1, 2, \ldots, m$, their difference may be denoted by $f_i = q_i - \hat{q}_i$, $i = 1, 2, \ldots, m$.

The objective functions applied in practice have assumed a variety of forms of which the following are typical:

(a) $$F = \sum |f_i|$$

(b) $$F = \sum f_i^2$$

(c) $$F = \sum (f_i^2)^k; \qquad k = 2, 3, \ldots$$

(d) $$F = \sum (f_i/q_i)^2$$

(e) $$F = \sum (f_i^2 q_i^k); \qquad k = 1, 2, \ldots$$

In each of the above definitions, the summation is taken over all $m$ pairs of observed and generated flows. Form (b) above, which involves minimising the sums of squares of the differences, has been employed extensively, although form (d) and form (e) are more useful when emphasis has to be placed on the reproduction of low and high flows respectively. For the comparative testing of models using different data sets, form (b) may be modified by writing:

(f) $$F' = \sqrt{(F/m)}/\bar{q}$$

where $\bar{q}$ is the mean of the $m$ values of $q_i$. The numerator of form (f) is the root mean square error (RSME) of the model synthesis. Alternatively, writing var $(q_i)$ as the variance of the $m$ values of $q$, another error criterion in the form of a coefficient of determination may be obtained:

(g) $$F'' = [\text{var}\,(q_i) - F/m]/\text{var}\,(q_i)$$

## 6.4. PRACTICAL APPLICATIONS

The methods of analysis and synthesis described above in Sections 6.2 and 6.3 may be shown to perform well on synthetic data generated from true linear systems. Unfortunately, recorded data are contaminated by errors which are introduced during measurement, recording and processing. The effect of such errors on the derivation of the response function, or TUH, of a catchment has received relatively little attention, the exception being the study by Laurenson and O'Donnell (1969) of the performance of the harmonic method, the Laguerre method, matrix inversion and the cascade of linear reservoirs (also referred to as the Nash cascade) on error-contaminated synthetic data. In this study, 3 hyetographs of effective rainfall and 2 shapes of TUH were employed to produce 6 'true' hydrographs of surface runoff. Errors where then introduced into the rainfall excess distribution, the runoff hydrograph, or both, and the chosen methods of analysis and synthesis applied to the erroneous data sets. The TUH so obtained was then compared with the original 'true' TUH.

The errors whose effects were studied included the following:

(1) the assumption of a uniform loss rate;
(2) overestimation of total rainfall volume;
(3) error in the rating curve of the gauging station at high stages;
(4) baseflow separation error;
(5) synchronisation between hyetograph and runoff hydrograph; and
(6) lack of synchronisation between individual raingauges.

Even when only one magnitude of error was investigated, this experiment involved 144 cases (2 TUH shapes by 3 hyetographs by 6 errors by 4 methods). In general, method of derivation, hyetograph shape and type of error were found to interact strongly. However, reasonable errors in total rainfall volume, rating curve extrapolation and baseflow separation all produced surprisingly low TUH errors. The errors in applying the Nash cascade to contaminated data were of the same order as those obtained when fitting 'true' data. The harmonic method performed the best in 7 out of 18 cases and was equal best in another 5, but was badly affected by the raingauge synchronisation error. For the harmonic, Laguerre and matrix methods, advancing the input by one time unit merely delayed the TUH by one time unit without affecting TUH shape.

A problem which is frequently encountered when methods of analysis are applied to observed rainfall and runoff records is the appearance of high-frequency oscillations in the derived TUH. Blank *et al.* (1971) have demonstrated that such oscillatory response functions can result from data errors, i.e. observations unrepresentative of the system, and are not necessarily attributable to non-linear catchment processes. Those authors suggested that the derived TUH should be smoothed by taking the averages of successive pairs of adjacent ordinates until a hydrograph-like shape was obtained. The use of low-pass digital filters applied to either the input, the input and the output, or the system response has been studied in detail by Rao and Delleur (1971) and Delleur and Rao (1971, 1974). During a major study of floods in the British Isles (Natural Environment Research Council, 1975), the matrix method was employed to derive unit hydrographs which were then smoothed by replacing each ordinate by the average of itself and its two neighbours, the operation being repeated twice.

The use of weighted and unweighted moving averages and low-pass filters as smoothing devices is unfortunately somewhat arbitrary. Moreover, if the analysis is performed by digital computer without operator intervention, the unnecessary smoothing of what would be an acceptable original derived TUH may distort the shape of the hydrograph and markedly reduce its peak ordinate. Such changes are particularly to be avoided when parameters of TUH shape appear as dependent variables in a regionalisation study. One of the principal advantages of the harmonic method for deriving the TUH is the direct connection between high-frequency oscillations and the magnitude of the high-order harmonic terms. A procedure for *controlled* rather than arbitrary smoothing of an oscillatory TUH based upon this connection has been outlined by Hall (1977).

The truncated harmonic method suggested by Hall (1977) is based upon the form of the linkage equations (6.11) by which the harmonic coefficients of order $j$ of the TUH depend only on the harmonic coefficients of the same order $j$ of the effective rainfall and surface runoff data. Individual harmonics may therefore be omitted from the series representation of the TUH without affecting the calculation or use of other $\alpha_j$, $\beta_j$. If therefore the ordinates of the TUH obtained by using all $p$ harmonics exhibit high-frequency oscillations, truncation at the high-order end of the series representation to (say) $q$ of the $p$ harmonics can assist in eliminating such oscillations. However, the amount of truncation that can be allowed is subject to the constraint that the hydrograph obtained by convolving the 'smoothed' TUH with the effective rainfall pattern should not depart significantly from the shape of the observed surface runoff hydrograph. Even when the TUH derived from all $p$ harmonics is of an acceptable shape, truncation allows a certain economy in the amount of computation. An objective procedure for determining the appropriate number of harmonics directly, based upon the use of the periodogram, has been outlined by Hall (1977). The periodogram of a set of ordinates, $y_i$, $i = 1, 2, \ldots, n$, consists of a plot of the function

$$I_j = \tfrac{1}{2}(A_j^2 + B_j^2) \tag{6.47}$$

against the angular frequency, $2\pi j/p$, where $A_0$, $A_j$, $B_j$, $j = 1, 2, \ldots, p$, are the harmonic coefficients of the series representation of the ordinates $y_i$. The $I_j$, $j = 1, 2, \ldots, p$, represent the contribution to the variance of the series $y_i$ provided by the $j$th harmonic (see Rayner, 1971).

In practice, hydrographs having a well-defined primary peak give rise to periodograms in which the $I_j$-values decrease rapidly with increasing angular frequency. Minimising the number of harmonics included in the series representation of the ordinates can therefore be equated to accepting as high a proportion of explained variance as possible for that number of harmonics. However, if interest is centred on minimising the number of harmonics included in the series representation of the TUH, subject to the fidelity with which the observed surface runoff hydrograph is reconstituted, the dependence of the $I_j$ on the corresponding individual periodogram ordinates of the effective rainfall distribution, $J_j$, and the TUH, $K_j$, is required. Forming the sums of squares of the coefficients $\alpha_j$, $\beta_j$ from eqns (6.11) leads after some algebra to the expression

$$I_j = (n^2/2)J_jK_j \tag{6.48}$$

Equation (6.48) shows that the proportion of the variance of the surface runoff hydrograph ordinates explained by the first $q$ harmonics, i.e. the sum of the $I_j$, $j = 1, 2, \ldots, q$, is directly proportional to the cumulative sum of the products $J_jK_j$, $j = 1, 2, \ldots, q$. The value of $q$ at which the explained variance exceeds a predetermined threshold therefore also establishes an appropriate number of harmonics at which to truncate the series representation of the TUH. The recommended procedure may therefore be summarised as follows:

(i) compute the variance of the observed surface runoff hydrograph ordinates, $y_i$, $i = 1, 2, \ldots, p$, and the periodogram ordinates $I_j$, $j = 1, 2, \ldots, p$, from the harmonic coefficients, $A_j$, $B_j$;
(ii) form cumulative sums of the $I_j$ and express each as a proportion of the variance of the $y_i$;
(iii) compare the values of these proportions with a predetermined

threshold of explained variance and find the smallest number of harmonics, $q$, which equals or exceeds this criterion;

(iv) compute the first $q$ harmonic coefficients $a_j$, $b_j$ of the effective rainfall distribution and form the first $q$ harmonic coefficients $\alpha_j$, $\beta_j$ of the TUH using the linkage equations; and

(v) compute the TUH ordinates from the $\alpha_j$, $\beta_j$, $j = 1, 2, \ldots, q$.

The major difficulty in applying the synthesis approach lies in the method of estimating the parameters of the conceptual model from the observed data. The larger the number of parameters, the greater the flexibility of the model in describing the data but the more onerous the fitting procedure. An experienced operator, familiar with the construction of a particular conceptual model, may be able to obtain a reasonable set of parameter values with a reasonable economy in time and effort. However, the same operator would almost certainly be less successful with another conceptual model, and a beginner can find the trial-and-error procedure a frustrating experience. Objective computer-based methods of estimating parameter values have therefore attracted considerable attention. Given a conceptual model with $n$ parameters, these so-called optimisation or 'hill-climbing' techniques in effect carry out a search in an $n$-dimensional vector space formed by $n$ mutually orthogonal axes until a point is found which maximises (or minimises) a predetermined objective function. The procedure which appears to be most commonly used is a modified version of that proposed by Rosenbrock (1960), which was recommended by Ibbitt and O'Donnell (1971). The latter authors have also provided a succinct review of the problems that may be encountered in applying optimisation techniques, and have drawn attention to the need to consider the characteristics of the method and its suitability in dealing with different shapes of response surface when choosing the structure of the conceptual model.

## REFERENCES


Amorocho, J. and Hart, W. E. (1964) A critique of current methods in hydrologic systems investigation. *Trans. Am. Geophys. Un.*, **45,** 307–21.

Blank, D., Delleur, J. W. and Giorgini, A. (1971) Oscillatory kernel functions in linear hydrologic models. *Wat. Resour. Res.*, **7,** 1102–17.

Dawdy, D. R. and O'Donnell, T. (1965) Mathematical models of catchment behaviour. *Proc. Am. Soc. Civ. Engrs., J. Hydraul. Div.*, **91**(HY4), 123–37.

De Laine, R. J. (1970) Deriving the unitgraph without using rainfall data. *J. Hydrol.*, **10,** 379–90.

Delleur, J. W. and Rao, A. R. (1971) Linear systems analysis in hydrology—the transform approach, the kernel oscillations and the effect of noise. In *Systems approach to hydrology*, Proc. 1st Bilateral US-Japan Seminar in Hydrology (Wat. Resour. Publ., Fort Collins) pp. 116–42.

Delleur, J. W. and Rao, A. R. (1974) Characteristics and filtering of noise in linear hydrologic systems. Int. Assoc. Hydrol. Sci., Publ. No. 101, pp. 570–9.

Diskin, M. H. (1977) On the derivation of linkage equations for Laguerre function coefficients. *J. Hydrol.*, **32,** 321–7.

Diskin, M. H. and Simon, E. (1977) A procedure for the selection of objective functions for hydrologic simulation models. *J. Hydrol.*, **34,** 129–49.

Dooge, J. C. I. (1968) The hydrologic cycle as a closed system. *Bull. Int. Assoc. Sci. Hydrol.*, **13**(1), 58–68.

Dooge, J. C. I. (1973) Linear theory of hydrologic systems. US Dept. Agric., Agric. Res. Serv., Tech. Bull. 1468, 327 pp.

Hall, M. J. (1977) On the smoothing of oscillations in finite-period unit hydrographs derived by the harmonic method. *Hydrol. Sci. Bull.*, **22,** 313–24.

Ibbitt, R. P. and O'Donnell, T. (1971) Fitting methods for conceptual catchment models. *Proc. Am. Soc. Civ. Engrs., J. Hydraul. Div.*, **97**(HY9), 1331–42.

Laurenson, E. M., and O'Donnell, T. (1969) Data error effects in unit hydrograph derivation. *Proc. Am. Soc. Civ. Engrs., J. Hydraul. Div.*, **95**(HY6), 1899–1917.

Nash, J. E. (1957) The form of the instantaneous unit hydrograph. Int. Assoc. Sci. Hydrol., Publ. No. 45, pp. 114–21.

Nash, J. E. (1960) A unit hydrograph study, with particular reference to British catchments. *Proc. Instn. Civ. Engrs.*, **17,** 249–82.

Natural Environment Research Council (1975) *Flood Studies Report*, Vol I: *Hydrological studies* (NERC, London) 550 pp.

O'Donnell, T. (1960) Instantaneous unit hydrograph derivation by harmonic analysis. Int. Assoc. Sci. Hydrol., Publ. No. 51, pp. 546–57.

O'Donnell, T. (1966) Methods of computation in hydrograph analysis and synthesis. In *Recent trends in hydrograph synthesis*, Proc. Tech. Mtg. No. 21, TNO, The Hague, pp. 65–102.

Rao, A. R. and Delleur, J. W. (1971) The instantaneous unit hydrograph: its calculation by the transform method and noise control by digital filtering. Purdue Univ., Wat. Resour. Res. Center, Tech. Rept. No. 20, 59 pp.

Rayner, J. N. (1971) *An introduction to spectral analysis.* Monographs in Spatial and Environmental Systems Analysis (Pion Ltd, London) 174 pp.

Rosenbrock, H. H. (1960) An automatic method of finding the greatest or least value of a function. *Computer J.*, **3,** 175–84.

Sarginson, E. J. and Bourne, D. E. (1969) The analysis of urban rainfall runoff and discharge. *J. Instn. Munic. Engrs.*, **96,** 81–5.

Sherman, L. K. (1932) Streamflow from rainfall by unit-graph method. *Engng. News-Record*, **108,** 501–5.

SNYDER, W. M. (1955) Hydrograph analysis by the method of least squares. *Proc. Am. Soc. Civ. Engrs.*, Separate No. 793, 25 pp.

SWINNERTON, C. J., HALL, M. J. and O'DONNELL, T. (1973) Conceptual model design for motorway stormwater drainage. *Civ. Engng. Pub. Wks. Rev.*, **68,** 123–9, 132.

TENNESSEE VALLEY AUTHORITY (1961) Matrix operations in hydrograph computations. TVA Office of Tributary Area Development, Res. Paper No. 1 (typescript).

VIESSMAN, W., KEATING, W. R. and SRINIVASA, K. N. (1970) Urban storm runoff relations. *Wat. Resour. Res.*, **6,** 275–9.

# 7

# Flood Routing

## 7.1. INTRODUCTION

The flow in an open channel may be conveniently classified according to the changes in depth which occur in both time and space. When the depth of flow remains constant over the time interval of interest, the discharge is said to be steady. If the depth changes with time, the flow is unsteady. Where the depth of flow is the same at all channel cross-sections, the discharge is said to be uniform. If the depth changes along the length of the channel, the flow is non-uniform or varied. The latter case may be further subdivided into rapidly and gradually varied flows according to whether or not the changes occur over a comparatively short distance. There are therefore six categories of flows to be considered (Chow, 1959):

| Steady flow | Unsteady flow |
|---|---|
| —uniform flow | —unsteady uniform flow |
| —varied flow | —unsteady (varied) flow |
| —gradually varied flow | —gradually varied unsteady flow |
| —rapidly varied flow | —rapidly varied unsteady flow |

Steady uniform flow is the most fundamental state treated in open channel hydraulics. In contrast, unsteady uniform flow would involve the water surface remaining parallel to the channel bed but fluctuating with time, a condition of little practical consequence. The term 'unsteady flow' is therefore applied exclusively to describe unsteady varied flow. Whereas rapidly varied flows are a local phenomenon, gradually varied flows are typified by shallow water surface slopes and small changes in depth between adjacent channel cross-sections that are observable over long distances. A ready example of the latter is provided by the movement of a flood wave down an open channel.

The basic partial differential equations describing gradually varied unsteady open channel flow, which are also referred to as the De Saint-Venant equations, consist of a continuity equation and a dynamic (or momentum) equation. The former may be established by considering the conservation of mass in an element of length $dx$ between two channel cross-sections (see Fig. 7.1). In an unsteady flow, the depth changes with time at a rate of $\partial y/\partial t$ and the discharge changes with distance at a rate of $\partial Q/\partial x$. Assuming that water is incompressible, the change in discharge through the element must equal the change in channel storage during any time increment $dt$. More formally,

$$(\partial Q/\partial x)\,dx\,dt + B\,dx\,(\partial y/\partial t)\,dt = 0 \qquad (7.1)$$

Since the discharge, $Q$, is the product of the mean flow velocity, $u$, and the area of flow, $A$, eqn (7.1) may be expanded to

$$A(\partial u/\partial x) + u(\partial A/\partial x) + B(\partial y/\partial t) = 0 \qquad (7.2)$$

Dividing through eqn (7.2) by $B$, and substituting the hydraulic mean depth, $R$, for the quotient $A/B$ and $\partial A = B\ \partial y$,

$$R(\partial u/\partial x) + u(\partial y/\partial x) + (\partial y/\partial t) = 0 \qquad (7.3)$$

The three terms in eqn (7.3) are referred to as the prism storage, the wedge storage and the storage caused by the rate of rise respectively.

The dynamic equation may be derived by applying the principle of the conservation of energy, whereby the overall change in total head,

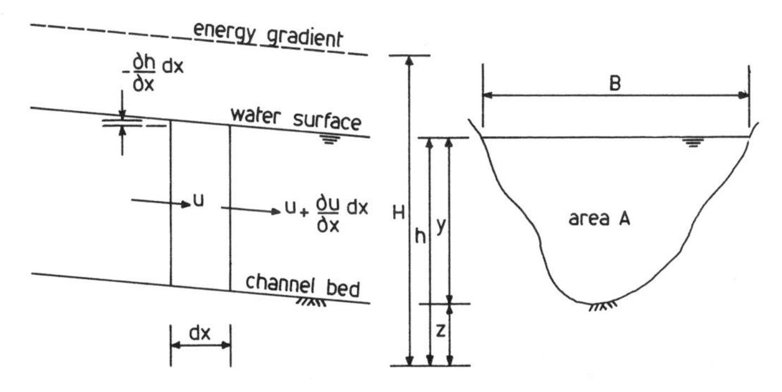


FIG. 7.1. Definition sketch for gradually varied unsteady flow.

$H$, within the distance d$x$ is equal to the loss due to the acceleration of the flow plus the loss due to frictional resistance. Applying Newton's second law of motion, the force due to the acceleration, $\partial u/\partial t$, acting upon a unit weight of water $W$ is given by $(W/g)(\partial u/\partial t)$, where g is the gravitational acceleration. The work done by this force over the length d$x$ is therefore $(W/g)(\partial u/\partial t)\,dx$ and the loss in head is $(1/g)(\partial u/\partial t)\,dx$. Denoting the friction slope by $S_f$, the dynamic equation may therefore be written as

$$dH = -(1/g)(\partial u/\partial t)\,dx - S_f\,dx \tag{7.4}$$

However, since the total head, $H$, is the sum of the velocity head, the pressure head and the elevation according to the Bernoulli equation,

$$H = (u^2/2g) + y + z \tag{7.5}$$

substituting for $H$ in eqn (7.4), dividing through by d$x$ and employing partial differentials leads to the expression

$$(1/g)(\partial u/\partial t) + (u/g)(\partial u/\partial x) + (\partial y/\partial x) + S_f - S_0 = 0 \tag{7.6}$$

where $S_0 = -(\partial z/\partial x)$ is the bed slope of the channel. The five terms in eqn (7.6) are referred to as the local acceleration, the convective acceleration, the pressure gradient, the friction slope and bed slope respectively.

Equations (7.2) and (7.6) are non-linear, and have no known analytical solution. However, given one initial and two boundary conditions, they may be solved by numerical methods for values of $y$ and $u$ at finite increments of $t$ and $x$. These methods may be applied either directly to eqns (7.2) and (7.6) or to the same expressions in an alternative form referred to as the characteristic equations. The latter are derived by considering a flood wave to be composed of a large number of infinitesimal surges, each of which has a discontinuous surface profile. At the point of discontinuity, $\partial y/\partial x$ has two values which are unrelated to each other, so that in mathematical terms the slope is indeterminate. This condition gives rise to the following equations (see Chow, 1959):

$$(dx/dt) = u \pm \surd(gy) \tag{7.7}$$

$$d[u \pm 2\surd(gy)] = g(S_0 - S_f)\,dt \tag{7.8}$$

Equations (7.7) and (7.8) are the characteristics which give the position of the discontinuities as they propagate in the $(x, t)$ plane. A comprehensive account of the method of characteristics has been presented by Abbott (1966). This approach has been applied to the solution of flood routing problems by Amein (1966), Fletcher and Hamilton (1967), Baltzer and Lai (1968), Mozayeny and Song (1969), Ellis (1970), Harris (1970), Wylie (1970), Pinkayan (1972), Jolly and Yevjevich (1974) and Sivaloganathan (1978a,b) among others.

Both explicit and implicit finite difference schemes (Brackensiek, 1967) are available for the solution of eqns (7.2) and (7.6). The former use explicit equations for $u$ and $y$ at given $x$, $t$ points, whereas the latter employ sets of simultaneous equations to find $u$ and $y$ at several values of $x$ but a single time point. Although explicit schemes are simpler to program, the size of the time step, $\Delta t$, is limited by the Courant stability condition:

$$\Delta t \leqslant \Delta x/[u+\sqrt{(gy)}] \tag{7.9}$$

where $\Delta x$ is the increment in distance. A variety of explicit schemes have been discussed by Yevjevich and Barnes (1970) and Sivaloganathan (1980). In contrast, implicit schemes are not subject to this stability condition, and are claimed to be more efficient for the majority of flood routing applications (see Amein, 1968; Amein and Fang, 1970; Chaudhry and Contractor, 1973; Halliwell and Ahmed, 1973; Amein and Chu, 1975; Price and Samuels, 1980).

Flood routing models based upon eqns (7.2) and (7.6) are generally referred to as complete dynamic models. Although such models provide a comprehensive description of the physical processes involved in gradually varied unsteady flow, they may prove to be very demanding in terms of computing resources. Moreover, their practical application is restricted by the assumptions inherent in the De Saint-Venant equations, among the more important of which are (Yevjevich, 1964):

(1) vertical velocities and accelerations are small in comparison with horizontal velocities and accelerations because of the gradual changes in discharge and depth with time and distance;
(2) the influence of the channel sides and their curvature can be neglected;
(3) the velocity distribution along a vertical is the same in steady and unsteady flow; and
(4) the frictional resistance is the same in steady and unsteady flow.

Bearing in mind the complex physical properties of natural river systems, even the complete dynamic models are restricted to long reaches and conditions under which the effects of channel irregularities are included within an empirical roughness coefficient. Therefore, since the application of eqns (7.2) and (7.6) may involve a conceptual element, the question arises as to whether more approximate models might produce results that are adequate for most practical purposes, if limited in their generality and accuracy, at considerably less expense. There are many different approximate models available but, as noted by Weinmann and Laurenson (1979), they are all essentially variants of the diffusion analogy and the kinematic wave models, the details of which are discussed in Section 7.2.

The passage of a flood wave through a reservoir may be regarded as a special case of open channel routing in which wedge storage is negligible, and eqns (7.2) and (7.3) reduce to the simple statement that the difference between the inflow and outflow is equal to the rate of change in the volume of storage. Moreover, if the water surface of the reservoir is effectively level, both the outflow discharge and the storage volume depend only upon its elevation. A storage–discharge relationship may therefore be readily obtained, which may be employed along with the continuity equation to derive the outflow hydrograph from the inflow hydrograph. Both tabular and graphical solution procedures are available, and are described in Section 7.3.

Although reservoir routing presents little numerical difficulty, the amount of computation can become excessive, particularly during the early stages in a design which involves the selection of the size of the outlet control structure required to limit the peak outflow to a specified maximum. If the proposed outlet control may be located at different elevations, the volume of the available storage also varies, thereby further complicating the exercise. In order that such design problems may be solved with a minimum expenditure of time and effort, a number of rapid routing procedures have been developed. Again, both numerical and graphical approaches can be applied, and examples of both are discussed in Section 7.4.

## 7.2. APPROXIMATE ROUTING METHODS

Following Weinmann and Laurenson (1979), the development of the approximate flood routing models is most conveniently illustrated by transforming eqn (7.6) into a rating curve, using the Chezy formula to describe the discharge, $Q_n$, of steady, uniform flow in an open channel:

$$Q_n = K\sqrt{(S_0)} \tag{7.10}$$

where $K$ is the channel conveyance. Since under the assumptions made in deriving eqns (7.2) and (7.6) the discharge for unsteady flow at the same depth is given by

$$Q = K\sqrt{(S_f)} \tag{7.11}$$

eqns (7.10) and (7.11) may be combined to give

$$Q = Q_n\sqrt{(S_f/S_0)} \tag{7.12}$$

Substituting for $S_f$ from eqn (7.6) then gives the general expression for the loop rating curve:

$$Q = Q_n\sqrt{\{1-(1/S_0)(\partial y/\partial x)-(u/gS_0)(\partial u/\partial x)-(1/gS_0)(\partial u/\partial t)\}} \tag{7.13}$$

The form of the loop rating curve is sketched in Fig. 7.2. The width of the loop depends upon the magnitude of the second, third and fourth terms within the braces in eqn (7.13). The different approximate models may also be identified according to the number of these terms which are retained. In particular, omission of the second, third and fourth terms leads to $Q = Q_n$ and a single-valued rating curve. Equation (7.6) assumes its simplest form:

$$S_f = S_0 \tag{7.14}$$

and the model may be completely defined by a single equation, which may be derived by rewriting eqn (7.2) as

$$(\partial Q/\partial x) + (\partial A/\partial Q)(\partial Q/\partial t) = 0 \tag{7.15}$$

Since the rating curve is single-valued, $\partial Q/\partial A = dQ/dA$, so that on evoking the Kleitz–Seddon principle and defining the celerity or travel

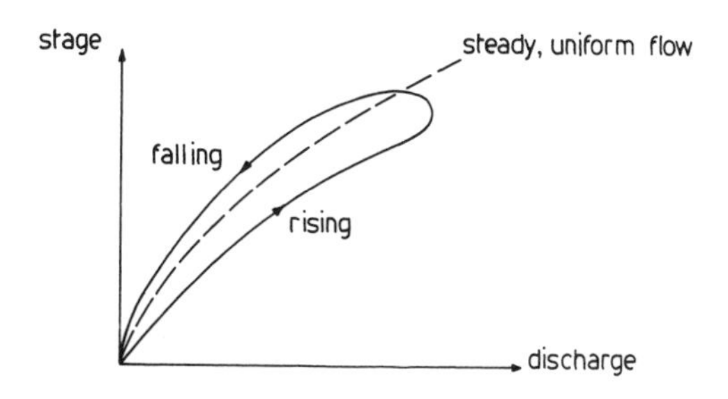


FIG. 7.2. The loop rating curve.

speed of the flood wave by

$$c = (dQ/dA) = (1/B)(dQ/dy) \tag{7.16}$$

eqn (7.15) becomes

$$(\partial Q/\partial t) + c(\partial Q/\partial x) = 0 \tag{7.17}$$

Equation (7.17) is referred to as the kinematic wave equation (see Lighthill and Whitham, 1955). Kinematic waves described by eqn (7.17) travel without attenuation, but with a change in shape, at the celerity given by eqn (7.16). However, the so-called kinematic models which are based upon eqn (7.16) are more general and account for attenuation through the numerical solution procedures that are applied. For problems with a very simple geometry, analytical solutions are available (see, for example, Henderson and Wooding, 1964). Kinematic models have also been widely employed as the routing component in hydrological catchment models (Wooding, 1965a,b, 1966; Woolhiser, 1969; Kibler and Woolhiser, 1970; Singh, 1975, 1976; Sherman and Singh, 1976a,b; Singh and Woolhiser, 1976).

A second important group of approximate models, which are referred to as the diffusion analogy models, may be characterised by omitting only the third and fourth terms in the equation for the loop rating curve. Equation (7.6) therefore reduces to

$$(\partial y/\partial x) + S_f - S_0 = 0 \tag{7.18}$$

If the continuity equation (7.2) written in the form

$$(\partial Q/\partial x) + B(\partial y/\partial t) = 0 \tag{7.19}$$

is differentiated with respect to $x$ and eqn (7.18) is differentiated with respect to $t$, and the results are combined,

$$\partial S_f/\partial t = (1/B)(\partial^2 Q/\partial x^2) \tag{7.20}$$

The left-hand side of eqn (7.20) may be expanded by writing the friction slope in terms of the discharge and channel conveyance according to eqn (7.11). After some algebraic manipulation (see Smith, 1980), eqn (7.20) reduces to

$$(\partial Q/\partial t) + c(\partial Q/\partial x) = D(\partial^2 Q/\partial x^2) \tag{7.21}$$

where $c$ is the flood wave celerity given by eqn (7.16) and $D$ is a diffusion coefficient defined by the expression

$$D = Q/(2BS_f) \tag{7.22}$$

When applied to flood routing problems, the coefficient $c$ accounts for the translation and the coefficient $D$ the attenuation of the wave. Equation (7.21) may be expressed in terms of depth of flow (Hayami, 1951; Thomas and Wormleaton, 1970) as well as discharge (Price, 1973; Natural Environment Research Council, 1975). Moreover, the two coefficients may either be selected for a representative discharge and held constant (Hayami, 1951) or expressed as functions of discharge (Natural Environment Research Council, 1975).

Equations (7.17) and (7.21) may be solved using the same numerical methods as those applied to the complete dynamic models. Using the kinematic model as an example, eqn (7.17) may be expressed in finite difference form as

$$(\Delta Q/\Delta t)+c(\Delta Q/\Delta x)=0 \tag{7.23}$$

Following Smith (1980), who employed weighting factors, $\alpha$ and $\beta$, in the space and time directions respectively to define the finite differences, as shown in Fig. 7.3:

$$\Delta Q/\Delta t=[\alpha(I_{m+1}-I_m)+(1-\alpha)(O_{m+1}-O_m)]/\Delta t \tag{7.24}$$

$$\Delta Q/\Delta x=[\beta(O_{m+1}-I_{m+1})+(1-\beta)(O_m-I_m)]/\Delta x \tag{7.25}$$

where $I_m$, $O_m$ are the inflow and outflow to a reach of length $\Delta x$ at time step $m$. Substituting eqns (7.24) and (7.25) into eqn (7.23) yields the following explicit expression for the outflow at time step $m+1$:

$$O_{m+1}=C_1I_{m+1}+C_2I_m+C_3O_m \tag{7.26}$$

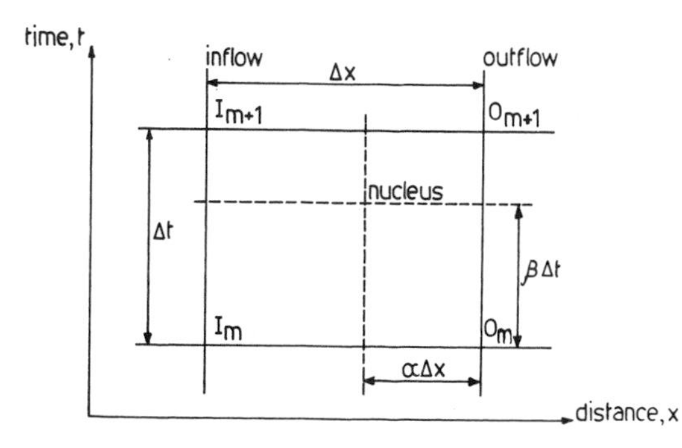


FIG. 7.3. Definition sketch for a weighted finite difference scheme in the $x$–$t$ plane.

where

$$C_1 = [c\beta\,\Delta t - \alpha\,\Delta x]/[c\beta\,\Delta t + (1-\alpha)\,\Delta x] \tag{7.27}$$

$$C_2 = [c(1-\beta)\,\Delta t + \alpha\,\Delta x]/[c\beta\,\Delta t + (1-\alpha)\,\Delta x] \tag{7.28}$$

$$C_3 = [(1-\alpha)\,\Delta x - c(1-\beta)\,\Delta t]/[c\beta\,\Delta t + (1-\alpha)\,\Delta x] \tag{7.29}$$

A recurrence relation identical in form to eqn (7.26) also features in another group of flood routing methods referred to as hydrological or storage routing models. These models do not incorporate flow resistance directly, but describe the influence of storage, $V$, within the channel reach. For this purpose, the continuity equation is rewritten in the form

$$(\mathrm{d}V/\mathrm{d}t) = I - O \tag{7.30}$$

the solution of which requires an expression relating the storage volume to the inflow and outflow. Perhaps the most widely known example of the latter is that originated by the US Army Corps of Engineers in 1938 in connection with the design of flood control works for the Muskingum river basin in Ohio:

$$V = w[\theta I + (1-\theta)O] \tag{7.31}$$

where $w$ is a storage constant and $\theta$ a weight, the values of which must be determined from observed inflow and outflow hydrographs. If eqn (7.30) is replaced by its finite difference equivalent:

$$(V_{m+1} - V_m)/\Delta t = [(I_{m+1} + I_m) - (O_{m+1} + O_m)]/2 \tag{7.32}$$

and the storages are expressed in terms of inflow and outflow according to eqn (7.31), the following equation for $O_{m+1}$ is obtained:

$$O_{m+1} = C_1' I_{m+1} + C_2' I_m + C_3' O_m \tag{7.33}$$

where

$$C_1' = (\Delta t - 2w\theta)/(2w(1-\theta) + \Delta t) \tag{7.34}$$

$$C_2' = (\Delta t + 2w\theta)/(2w(1-\theta) + \Delta t) \tag{7.35}$$

$$C_3' = (2w(1-\theta) - \Delta t)/(2w(1-\theta) + \Delta t) \tag{7.36}$$

Equations (7.33)–(7.36) constitute the Muskingum method of flood routing. A comparison between the coefficients of eqns (7.26) and (7.33) shows immediately that eqns (7.27)–(7.29) are identical to eqns (7.34)–(7.36) if $\alpha = 0$, $\beta = 0{\cdot}5$ and $w = \Delta x/c$. The Muskingum method is therefore a particular case of the kinematic model.

The conventional method of determining appropriate values of $w$

and $\theta$ from observed hydrographs involves a trial-and-error procedure in which a value of $\theta$ is assumed and a plot prepared of the weighted flow, $\theta I_m + (1-\theta)O_m$, against the storage at each time step $m$. In practice, the changes in storage at each time step are computed from eqn (7.32) and cumulative sums of these changes formed for successive time intervals. Such graphs of storage against weighted flow generally plot in the form of a loop. The value of $\theta$ is then adjusted so that the width of the loop is minimised. $w$ is obtained from the gradient of the line of best fit through the plotted points. Strictly, this line should also pass through the origin of the graph. The application of the above procedure may prove to be particularly difficult in certain cases owing to the shape of the loops in the graph of weighted flow against storage. However, a more objective approach in which optimum values of $w$ and $\theta$ (in the least-squares sense) may be computed directly has been suggested by Gill (1978).

As expressed in eqn (7.31), $V$ is an absolute value. In practice, estimates of storage are relative, so that a more general form of equation, such as

$$V = w[\theta I + (1-\theta)O] + \epsilon \tag{7.37}$$

is to be preferred, where $\epsilon$ is the difference between relative and absolute storages. By writing

$$E = w\theta \tag{7.38}$$

$$F = w(1-\theta) \tag{7.39}$$

eqn (7.37) simplifies to

$$V = EI + FO + \epsilon \tag{7.40}$$

The problem of determining the values of $E$, $F$ and $\epsilon$ that minimise the width of the weighted flow/storage loop may therefore be expressed in an alternative form by defining

$$\delta_m = V_m - (EI_m + FO_m + \epsilon) \tag{7.41}$$

and searching for the values of $E$, $F$ and $\epsilon$ that minimise the sums of squares of the $\delta_m$. The application of standard least-squares procedures results in the following normal equations:

$$\sum V - E\sum I - F\sum O - n\epsilon = 0 \tag{7.42}$$

$$\sum VI - E\sum I^2 - F\sum IO - \epsilon\sum I = 0 \tag{7.43}$$

$$\sum VO - E\sum IO - F\sum O^2 - \epsilon\sum O = 0 \tag{7.44}$$

where all summations are taken over the $n$ values of $V$, $I$ and $O$. Once eqns (7.42)–(7.44) have been solved for $E$, $F$ and $\epsilon$,

$$w = E + F \tag{7.45}$$

$$\theta = E/w \tag{7.46}$$

The results of applying this procedure are illustrated in Fig. 7.4 for a flood which occurred in March 1979 on the Kali Pemali, a river in north-central Java. The distance between the upstream (Rengaspendawa) and downstream (Brebes) gauging stations was approximately 14 km. The computed hydrograph at Brebes, which was obtained using eqn (7.33) with values of $w = 4{\cdot}036$ and $\theta = 0{\cdot}164$ in eqns (7.34)–(7.36), agrees well with the observed hydrograph, as shown in Fig. 7.4. A more refined method of determining optimum values of the coeffi-

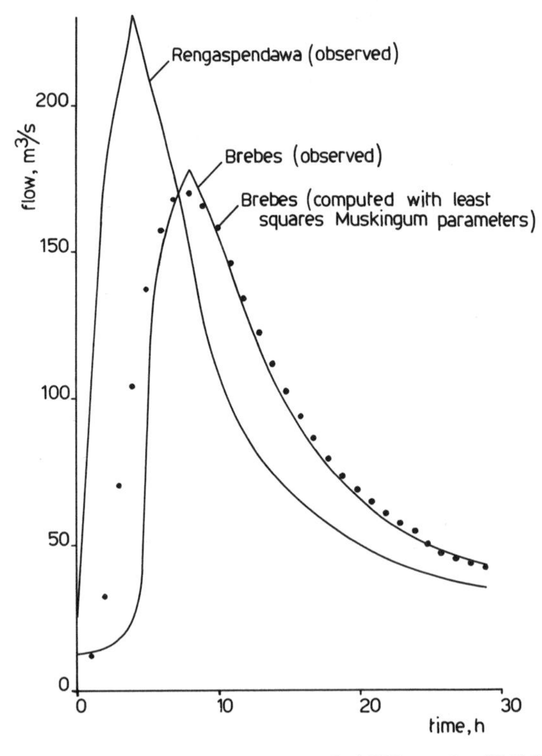


FIG. 7.4. Routing of the flood of 26–27 March 1979 on the Kali Pemali using the Muskingum method.

cients $w$ and $\theta$, using a mathematical programming technique, has been described by Stephenson (1979).

Figure 7.4 provides a ready illustration that, despite the basic underlying assumption of the kinematic model and the Muskingum method that the rating curve is single-valued, the latter approach is capable of simulating flood wave attenuation. The apparent anomaly may be directly attributed to the replacement of the partial differentials in the continuity equation by finite differences. As demonstrated by both Cunge (1969) and Smith (1980), the finite difference scheme introduces an error into the analytical solution of the original equation in the form of an artificial attenuation of the flow rate and a modification to the flood wave celerity. By expanding each finite difference term as a Taylor series about the nucleus of the grid at which the differential equation is computed (see Fig. 7.3), those authors also showed that eqn (7.33) is a finite difference approximation to the convective diffusion equation (7.21) as well as the kinematic wave equation (7.17). An alternative (and simpler) demonstration of the same result has been presented by Koussis (1978), who employed the original eqns (7.30) and (7.31) instead of the numerical solution scheme. The derivation is based upon the expansion of the outflow discharge as a Taylor series in $x$, i.e.

$$O = I + \Delta x I' + (\Delta x^2/2) I'' + \cdots \tag{7.47}$$

where $I'$ and $I''$ denote the first and second derivatives of the inflow discharge to the reach of length $\Delta x$. This expression is then substituted into eqns (7.30) and (7.31), and the latter differentiated with respect to time. The two expressions for $\mathrm{d}V/\mathrm{d}t$ so obtained are equated, and after some algebraic manipulation involving eqns (7.16) and (7.17),

$$(\partial Q/\partial t) + (\Delta x/w)(\partial Q/\partial x) = [\Delta x(1-\theta)c - (\Delta x^2/2w)](\partial^2 Q/\partial x^2) \tag{7.48}$$

where the more general symbol $Q$ has been substituted for the inflow discharge. Equation (7.48) has the general form of the convective diffusion equation (7.21). Moreover, equating coefficients, the following expressions are obtained for the constant $w$ and the weight $\theta$:

$$w = \Delta x/c \tag{7.49}$$

$$\theta = [1 - (Q/BS_f c\, \Delta x)]/2 \tag{7.50}$$

Equations (7.49) and (7.50), which were first derived by Cunge (1969), are of particular importance in enabling values of the Muskingum coefficients to be computed from physical properties of the channel reach and the wave celerity at a representative discharge. This approach, which has become known as the Muskingum–Cunge method, has been applied to flood routing problems in the United Kingdom by Price (1978) and the Natural Environment Research Council (1975). A variable parameter Muskingum–Cunge method has been described by Price (1978) and Ponce and Yevjevich (1978).

As presented above, both the Muskingum and Muskingum–Cunge methods involve the estimation of the coefficients, $w$ and $\theta$, required to route an inflow hydrograph, given $\Delta t$ and the length of the channel reach. (In eqns (7.49) and (7.50), the reach length, $L$, is assumed equal to the space scale, $\Delta x$.) Alternatively, one (or both) of the coefficients might be specified, and either the reach length or both the time and space scales estimated for the finite difference scheme. For example, in the Kalinin–Miljukov method (Dooge, 1973), the weight $\theta$ is set to zero, and the reach of interest is divided into characteristic lengths, each of which is treated as a linear storage element. Under these conditions, eqn (7.50) shows that the length of each sub-reach, $\Delta x$, is given by

$$\Delta x = Q/(BS_f c) \tag{7.51}$$

Ponce (1979) has suggested another simplified routing method in which $\theta$ is set to zero and the storage constant, $w$, is equated to the time step, $\Delta t$. Equations (7.34)–(7.36) then reduce to

$$C_1' = C_2' = C_3' = 1/3 \tag{7.52}$$

In addition, the length of each sub-reach is given by eqn (7.51). Assuming that for the channel reach of interest, discharge is related to the cross-sectional area of flow by the equation

$$Q = aA^b \tag{7.53}$$

where $a$ and $b$ are constants, the wave celerity may be computed from eqn (7.16) as

$$c = bQ/A \tag{7.54}$$

Substituting eqn (7.54) in eqns (7.51) and (7.49),

$$\Delta x = A/(bS_f B) \tag{7.55}$$

$$\Delta t = A^{2-b}/(S_f Bab^2) \tag{7.56}$$

Equations (7.55) and (7.56) may therefore be used to estimate the required space and time scales, given values of $A$ and $B$ at a representative discharge, the discharge–area relationship of eqn (7.53) and an estimate of the friction slope. The ordinates of the inflow hydrograph are then read off at intervals of $\Delta t$, and then routed $L/\Delta x$ times using the recurrence relation

$$O_{m+1} = (I_{m+1} + I_m + O_m)/3 \tag{7.57}$$

in order to produce the required outflow hydrograph. General guidance on the choice of $\Delta x$ and $\Delta t$ that will ensure the greatest accuracy when applying the Muskingum–Cunge method has been provided by Jones (1981).

## 7.3. FULL RESERVOIR ROUTING METHODS

As noted in Section 7.1, reservoir routing may be regarded as a particular case of open channel flood routing whose solution is greatly simplified by the storage being a function only of the reservoir outflow. The latter condition arises from both the volume of storage in the reservoir and the outflow discharge being a function of water surface elevation, and the application of the assumption that the surface of the reservoir remains effectively horizontal. The frequent use of the term ‘level pool routing’ to describe reservoir routing is sufficient to emphasise the central importance of this assumption in the standard solution procedures that are described in this section.

Reservoirs and flood storage ponds may be broadly classified as being either onstream or offstream. With the former, dry weather flow passes through the storage area, whereas with the latter, the storage is bypassed. The methods outlined below apply specifically to onstream reservoirs, which have a single control structure at their outlet. In contrast, offstream reservoirs generally have control structures which regulate both the inflow to and the outflow from storage; many have a third structure located in the watercourse from which diversion takes place between the entry to and the exit from the reservoir. Offstream reservoirs and ponds are therefore more conveniently analysed by extending the methods applicable to onstream reservoirs according to the particular configuration of the site.

Since the reservoir storage–discharge relationship may be highly non-linear, the full routing procedures are based upon the finite difference form of the continuity equation given by eqn (7.32), with the inflow hydrograph provided in the form of a series of ordinates, $I_m$, at equally spaced time intervals of $\Delta t$. Rearranging eqn (7.32) so that all known terms are on the left-hand side gives

$$(I_m + I_{m+1})/2 + [(V_m/\Delta t) - (O_m/2)] = [(V_{m+1}/\Delta t) + (O_{m+1}/2)] \quad (7.58)$$

The routing procedure therefore consists of substituting the known values of $I_m$, $I_{m+1}$, $V_m$ and $O_m$ into eqn (7.58) to obtain $(V_{m+1}/\Delta t)+(O_{m+1}/2)$, and the determination of $O_{m+1}$ using the storage–discharge relationship of the reservoir. The complete outflow hydrograph is produced by the recursive solution of eqn (7.58) using either tabular or graphical methods. The application of such procedures is most conveniently demonstrated by means of a worked example.

Table 7.1 illustrates the application of the tabular method to route an inflow hydrograph through an onstream pond having an outlet control consisting of a broad-crested weir, 3·95 m in length, having a discharge coefficient of 2·1. Its rating is therefore given by the expression

$$O = 8{\cdot}30h^{1{\cdot}5}$$

where $h$ is the head above the weir (m), and the outflow discharge is expressed in $m^3/s$. When combined with the storage–height curve of the reservoir, given by

$$V = 112\,000h^{1{\cdot}25}$$

where the storage is expressed in $m^3$, the following storage–discharge relationship is obtained:

$$V = 19\,337O^{0{\cdot}83}$$

The ordinates of the inflow hydrograph are entered into column 2 of Table 7.1, and the corresponding times into column 1. In this example, $\Delta t$ is taken as 1800 s. The ordinates in column 3 are obtained by averaging successive pairs of ordinates from column 2. The ordinates of the outflow hydrograph appear in column 4. In order to initiate the routing, the first ordinate must be provided. A value of 0·1% of the peak inflow rate gives a reasonable starting figure.

Given $O_1$, the term $(V_1/\Delta t)-(O_1/2)$ may be computed (with $\Delta t$ expressed in compatible units) and entered into column 5. The entries in columns 3 and 5 now provide the two terms on the left-hand side of eqn (7.58), and their sum, equal to $(V_2/\Delta t)+(O_2/2)$, is entered into

**Table 7.1. Example of the tabular full reservoir routing method** (modified from Hall and Hockin, 1980, by permission of CIRIA)

| (1) | (2) | (3) | (4) | (5) | (6) |
|---|---|---|---|---|---|
| *Time* | $I$ ($m^3/s$) | $\frac{I_m + I_{m+1}}{2}$ | $O$ ($m^3/s$) | $\frac{V_m}{\Delta t} - \frac{O_m}{2}$ | $\frac{V_{m+1}}{\Delta t} + \frac{O_{m+1}}{2}$ |
| 0·0 | 0·41 | 0·48 | 0·022 | 0·44 | 0·92 |
| 0·5 | 0·54 | 0·74 | 0·05 | 0·87 | 1·61 |
| 1·0 | 0·94 | 1·31 | 0·10 | 1·51 | 2·82 |
| 1·5 | 1·68 | 2·26 | 0·19 | 2·63 | 4·88 |
| 2·0 | 2·82 | 3·80 | 0·37 | 4·51 | 8·31 |
| 2·5 | 4·77 | 6·54 | 0·70 | 7·61 | 14·15 |
| 3·0 | 8·31 | 10·68 | 1·32 | 12·83 | 23·51 |
| 3·5 | 13·04 | 15·26 | 2·41 | 21·10 | 36·36 |
| 4·0 | 17·48 | 18·96 | 4·05 | 32·31 | 51·27 |
| 4·5 | 20·44 | 21·02 | 6·10 | 45·17 | 66·19 |
| 5·0 | 21·60 | 21·32 | 8·27 | 57·92 | 79·24 |
| 5·5 | 21·03 | 20·19 | 10·25 | 68·99 | 89·18 |
| 6·0 | 19·34 | 18·14 | 11·79 | 77·39 | 95·53 |
| 6·5 | 16·93 | 15·53 | 12·80 | 82·73 | 98·26 |
| 7·0 | 14·13 | 12·80 | 13·23 | 85·03 | 97·83 |
| 7·5 | 11·46 | 10·26 | 13·16 | 84·67 | 94·93 |
| 8·0 | 9·06 | 8·08 | 12·70 | 82·23 | 90·31 |
| 8·5 | 7·09 | 6·32 | 11·97 | 78·34 | 84·66 |
| 9·0 | 5·54 | 4·95 | 11·09 | 73·57 | 78·52 |
| 9·5 | 4·35 | 3·90 | 10·14 | 68·38 | 72·28 |
| 10·0 | 3·45 | 3·11 | 9·19 | 63·09 | 66·20 |
| 10·5 | 2·76 | | | | |

column 6. The storage–discharge relationship is then employed to find the corresponding value of $O_2$. For convenience, a separate graph showing the variation of $(V/\Delta t)+(O/2)$ with outflow discharge may be prepared from the storage–discharge relationship. The value of $O_2$ is entered in the next row of column 4. The procedure is then repeated, with the value of $(V/\Delta t)-(O/2)$ being obtained by subtracting the new outflow rate from the value of $(V/\Delta t)+(O/2)$ in column 6 at the previous time step.

Alternatively, the above example may be approached using a graphical rather than a tabular routing procedure. Of the former, the method proposed by Sorensen (1952) appears to be among the most widely used (see also Cornish, 1974), although the 'Z' diagram described by

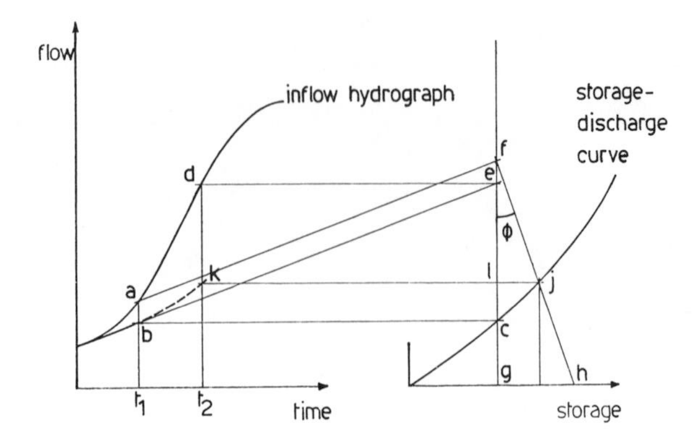


FIG. 7.5. Construction for the graphical full routing method.

Blackmore (1952) provides a more compact presentation. In the Sorensen method, both the inflow hydrograph and the storage–discharge relationship of the reservoir are plotted side by side on a large sheet of drawing paper with their discharge ordinates to the same scale, as shown in Fig. 7.5. The outflow hydrograph is then constructed according to the following procedure:

(i) At time $t_1$, which is located shortly after the beginning of the rise of the inflow hydrograph, a vertical line is drawn to intersect the inflow hydrograph at point $a$ at which the discharge is $I_1$.

(ii) An initial outflow discharge, $O_1$, at time $t_1$ is assumed which exceeds the initial flow but is less than $I_1$, as given by the point $b$; the storage, $V_1$, corresponding to $O_1$ is obtained by drawing the line $bc$ parallel to the time and discharge axes to intersect the storage–discharge curve at the point $c$.

(iii) At time $t_2$, which is located at a suitable time interval after $t_1$, a vertical line is drawn to intersect the inflow hydrograph at the point $d$ at which the discharge is $I_2$; the line $de$ is then drawn parallel to the time and storage axes to intersect a vertical through the point $c$ at the point $e$.

(iv) A line is drawn to connect the points $b$ and $e$, and a second line is drawn through the point $a$ parallel to the line $be$ to meet the vertical through the point $c$ at the point $f$.

(v) Denoting the intersection of the vertical through the point $c$ with the storage axis as the point g, the point $h$ on the storage axis is found for which $gh = (t_2 - t_1)(fg/2)$, i.e. the angle $\phi = \tan^{-1} 0{\cdot}5(t_2 - t_1)$.

(vi) From the point $j$ at which $fh$ intersects the storage–discharge curve, the line $jk$ is drawn parallel to the time and storage axes to meet the vertical through the point $d$ at the point $k$; the required value of the outflow discharge, $O_2$, at time $t_2$ is then given by the point $k$.

(vii) Steps (iii) to (vi) are repeated in order to obtain further points on the outflow hydrograph.

The proof of the construction involved in step (v) may be obtained by noting from Fig. 7.5 that

$$fg = ge + ef = I_2 + (I_1 - O_1)$$

By construction, $jl = 0{\cdot}5\ \Delta t\, fl$, where $\Delta t = t_2 - t_1$, and since $fl = fg - gl$,

$$jl = 0{\cdot}5\ \Delta t\ (I_1 + I_2 - O_1 - O_2)$$

Since $jl = V_2 - V_1$, this equation may be rearranged to give

$$(I_1 + I_2)/2 = (O_1 + O_2)/2 + (V_2 - V_1)/\Delta t$$

which is equivalent in form to the continuity equation as given by eqns (7.32) and (7.58).

The 'Z' diagram approach suggested by Blackmore (1952) differs from the Sorensen method in that the plot of the inflow hydrograph is turned through 90° and superimposed upon the plot of the storage–discharge curve, as shown in Fig. 7.6. The starting point 1 for the routing is defined by the intersection of the vertical drawn from the initial inflow at the point 1′ and the storage–discharge curve. From the point 1, a line is drawn at an angle of $\tan^{-1}(\Delta t/2)$, where $\Delta t$ is the time interval for the routing, to intersect the vertical drawn through the point 2′ on the inflow hydrograph at the end of the first time increment. From this intersection point a second line is taken back across the diagram at the same angle to meet the storage–discharge curve at the point 2, which defines the ordinate of the outflow hydrograph at the end of the first time increment. This construction is then repeated to give further points on the outflow hydrograph.

The proof of this construction may be demonstrated by referring to the inset on Fig. 7.6 showing the geometry of the first 'Z' line in more detail. In this sketch,

$$V_2 - V_1 = \Delta V = (I_2 - O_1) \tan \xi + (I_2 - O_2) \tan \xi$$

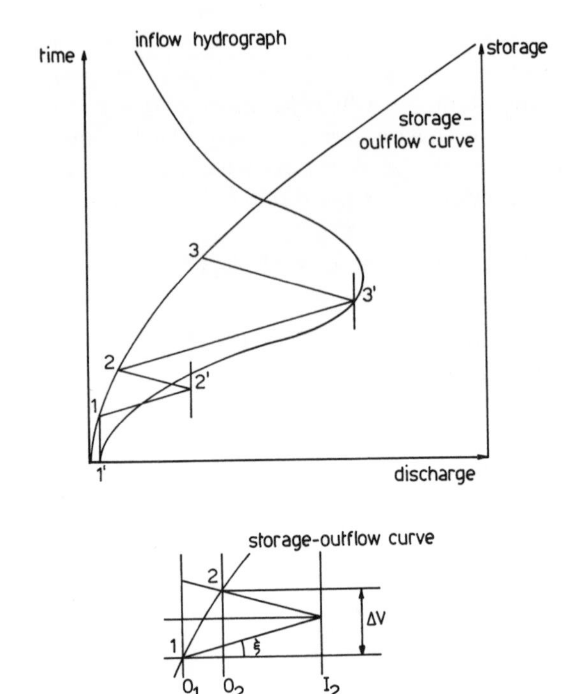


FIG. 7.6. Construction for the Blackmore graphical routing method.

and

$$\tan \xi = \Delta V/(2I_2 - O_1 - O_2)$$

However, the denominator in this expression for $\tan \xi$ is equal to twice the difference between the inflow and the average outflow during the second time increment, which is equivalent to twice the rate of change in storage, $2\,\Delta V/\Delta t$. The tangent of the angle $\xi$ is therefore $\Delta t/2$, as assumed in the construction.

## 7.4. RAPID RESERVOIR ROUTING METHODS

Rapid reservoir routing techniques are primarily intended to provide a convenient and economical method of screening alternatives at an early stage in a design exercise. Unlike the full routing procedures described in Section 7.3, the rapid methods are capable of producing only estimates of the peak outflow discharge, given the available storage volume, or estimates of the storage required to cause a predetermined level of attenuation of the peak inflow rate. Once the number of alternative design configurations has been reduced to two or possibly three, the final choice may be made on the basis of the more detailed results obtainable by applying a full routing procedure.

The majority of the rapid routing methods depend upon the use of geometrical figures, such as a triangle (Wycoff and Singh, 1976; Colombi and Hall, 1977) or a trapezium (West, 1974), or mathematical functions, such as the gamma distribution (Sarginson, 1973), to approximate the shape of the inflow hydrograph. With the former, the relationship between the peak rates of inflow and outflow and the active volume of storage follows directly from the geometry of the approximation. With the latter, the functional form allows the routing to be expressed in terms of a dimensionless differential equation. This equation may then be solved by numerical methods to give tables or graphs of functions from which the maximum outflow (or the storage volume) may be computed directly. This approach may be readily illustrated by a generalised and simplified version of the method proposed by Sarginson (1973) in which both the storage volume and the outflow discharge of the reservoir are expressed as simple power functions of the head above the outlet control (see Hall and Hockin, 1980).

The three-parameter gamma distribution was employed by Mitchell (1962) to approximate the variation of stage with time at the outlet of a reservoir. The same function is used here to describe the shape of the inflow hydrograph:

$$I = I_0(t/\gamma)^m \exp(-t/\gamma) \tag{7.59}$$

where $I$ is the inflow discharge at time $t$, and $I_0$, $\gamma$ and $m$ are parameters of the function.

Differentiating eqn (7.59) with respect to $t$ and equating the result to zero shows that the time-to-peak, $T_p$, of this inflow hydrograph is given by

$$T_p = \gamma m \tag{7.60}$$

Substituting this value of $T_p$ into eqn (7.59) gives the peak discharge:

$$I_p = I_0 m^m \exp(-m) \tag{7.61}$$

Eliminating $I_0$ from eqns (7.59) and (7.61) then provides the following alternative expression for the functional form of the inflow hydrograph:

$$I = I_p[t/(\gamma m)]^m \exp[m - (t/\gamma)] \tag{7.62}$$

In addition to adopting eqn (7.62) to describe the shape of the inflow hydrograph, the variation of the reservoir storage volume, $V$, with the height of the water surface above the outlet control, $h$, may be assumed to take the form

$$V = V_0 h^n \tag{7.63}$$

where $V_0$ and $n$ are a constant and exponent respectively, the former being the storage at a height of 1 m above the outlet control. In order to obtain the storage–discharge relationship for the reservoir, the further assumption is made that the rating curve of the outlet control may be described by the expression

$$O = kh^r \tag{7.64}$$

where $O$ is the outflow discharge, and $k$ and $r$ are a constant and exponent respectively. Substituting for $h$ from eqn (7.64) into eqn (7.63) gives the required storage–discharge relation:

$$V = V_0(O/k)^{n/r} \tag{7.65}$$

which, by writing $\upsilon = n/r$ and $C = V_0/k^\upsilon$, may be simplified to give the familiar 'non-linear reservoir' equation:

$$V = CO^\upsilon \tag{7.66}$$

The inflow, outflow and rate of change of storage in the reservoir may now be linked by means of the continuity equation. Substituting for $I$ from eqn (7.62) and computing $dV/dt$ from eqn (7.66) leads to the expression

$$(O + \upsilon CO^{\upsilon-1})\,dO/dt = I_p[t/(\gamma m)]^m \exp[m - (t/\gamma)] \tag{7.67}$$

This equation may be simplified by combining the five constants and two variables into the dimensionless groups

$$X = t/\gamma; \qquad Z = O/I_p; \qquad \Omega = CI_p^{\upsilon-1}/\gamma$$

which, when substituted into eqn (7.67), give

$$(Z + \upsilon)\Omega Z^{\upsilon-1}\,dZ/dX = (X/m)^m \exp(m - X) \tag{7.68}$$

This non-linear differential equation may be solved by numerical methods using a digital computer to give the outflow hydrograph as a proportion of the peak of the inflow hydrograph. In the special case of $v = 1$ (corresponding to a linear storage–discharge relationship), eqn (7.68) may be solved analytically. However, the expression for $Z$ involves the incomplete gamma function, and the numerical solution of the latter is no simpler than that of eqn (7.68).

When applied to a reservoir design problem, only the peak outflow, $O_p$, is required, since the maximum head and the maximum volume of storage may be obtained from $O_p$ by inverting eqns (7.64) and (7.63) respectively. Equation (7.68) may be solved for a range of the variables, $\Omega$, $m$ and $v$, which are likely to be encountered in practice to give, at the maximum of $Z$, values of the discharge attenuation ratio

$$R_q = O_p/I_p \tag{7.69}$$

$R_q$ may be represented as a series of curves against $m$ for various values of $\Omega$ and constant values of $v$. A comprehensive selection of such graphs has been provided by Hall and Hockin (1980).

In order to obtain the time-to-peak of the outflow hydrograph, $t_p$, eqn (7.69) may be expanded by noting that when $t = t_p$, $I = O_p$. Using eqn (7.59) and defining the timing ratio, $R_t = t_p/T_p$, eqn (7.69) becomes after some simplification

$$R_q^{1/m} = R_t \exp(1 - R_t) \tag{7.70}$$

Once $R_q$ is obtained, the timing ratio, and therefore $t_p$, may be determined from a plot of $R_t$ against $R_t \exp(1 - R_t)$.

The application of this numerical rapid routing method to a design problem requires a technique for fitting eqn (7.62) to the inflow hydrograph, and a procedure for fitting eqn (7.63) to the information obtained from a site survey. Of the several techniques available for fitting a gamma distribution to the inflow hydrograph, the three-point method described by Sarginson (1969) has the merit of simplicity commensurate with a rapid routing procedure. In this method, the discharges at the peak, $I_p$, and at 0·5 and 1·5 times the time-to-peak, $I_1$ and $I_2$, are read from a plot of the inflow hydrograph and the following equation solved for $m$:

$$I_p^2/(I_1 I_2) = (4/3)^m \tag{7.71}$$

Given this value of $m$, $\gamma$ may be computed from eqn (7.60).

With regard to the storage–height relationship, water surface areas at different heights above the crest of the outlet control may be measured

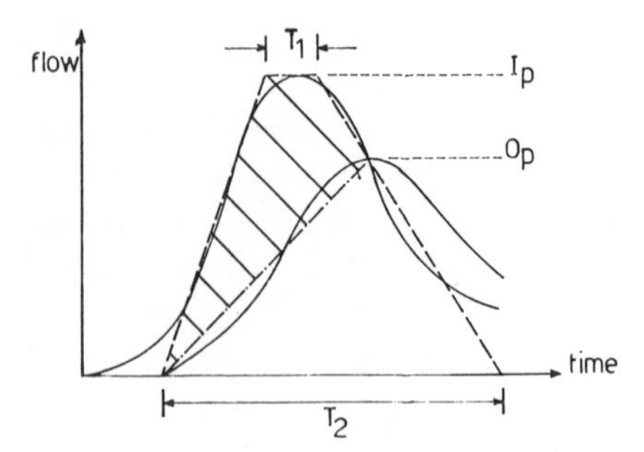


FIG. 7.7. Trapezoidal approximation to the inflow hydrograph used in a graphical rapid routing method.

from a detailed survey map of the pond site using a planimeter. Storage may then be estimated using (say) the prismatic rule. The gradient of a double-logarithmic plot of storage against height determines the exponent $n$, and $V_1$ may be read directly from the same graph at $h=1$.

Although the derivation of eqn (7.68) is somewhat protracted, the computed functions provide the 'exact' solution to the routing of the fitted gamma distribution through a pond with a specified storage–discharge curve within the limits of accuracy of the numerical solution procedure. No further approximations are made in the production of these functions. In contrast, the rapid routing methods which depend upon geometrical approximations to the shape of the inflow hydrograph also assume a linear rising limb to the outflow hydrograph. A typical example of the latter methods is provided by the approach suggested by West (1974), in which the inflow hydrograph is approximated by a trapezium, as shown in Fig. 7.7. The storage required,

$$V = I_p(T_1 + T_2)/2 - O_p T_2/2 \tag{7.72}$$

Using this value of $V$, eqn (7.63) may be inverted to give the maximum head on the outlet control:

$$h_p = (V_j/V_0)^{1/n} \tag{7.73}$$

Given both $O_p$ and $h_p$, the dimensions of the control may be obtained from the value of $k$ computed by inverting eqn (7.64).

Alternatively, if the amount of attenuation corresponding to a given volume of storage is required, eqn (7.72) may be rearranged to give

$$O_p = I_p[1 + (T_1/T_2)] - 2V/T_2 \tag{7.74}$$

Given $V$, $h_p$ may be obtained from eqn (7.73) and the dimensions of the control follow from eqn (7.64).

## REFERENCES

Abbott, M. B. (1966) *An introduction to the method of characteristics* (Thames & Hudson, London) 243 pp.

Amein, M. (1966) Streamflow routing on computer by characteristics. *Wat. Resour. Res.*, **2,** 123–30.

Amein, M. (1968) An implicit method for numerical flood routing. *Wat. Resour. Res.*, **4,** 719–26.

Amein, M. and Chu, H.-L. (1975) Implicit numerical modelling of unsteady flows. *Proc. Am. Soc. Civ. Engrs., J. Hydraul. Div.*, **101**(HY6), 717–31.

Amein, M. and Fang, C. S. (1970) Implicit routing in natural channels. *Proc. Am. Soc. Civ. Engrs., J. Hydraul. Div.*, **96**(HY12), 2481–2500.

Baltzer, R. A. and Lai, C. (1968) Computer simulation of unsteady flows in waterways. *Proc. Am. Soc. Civ. Engrs., J. Hydraul. Div.*, **94**(HY4), 1083–1117.

Blackmore, W. E. (1952) Line diagrams for problems of storage. *Wat. Power*, **4,** 299–303, 354–7.

Brackensiek, D. L. (1967) Finite difference methods. *Wat. Resour. Res.*, **3,** 847–60.

Chaudhry, Y. M. and Contractor, D. N. (1973) Application of implicit method to surges in open channels. *Wat. Resour. Res.*, **9,** 1605–12.

Chow, V. T. (1959) *Open-channel hydraulics* (McGraw-Hill, New York) 680 pp.

Colombi, J. S. and Hall, M. J. (1977) A quick screening method for estimating the routing effect of a reservoir. *Proc. Instn. Civ. Engrs. Part 2*, **63,** 935–41.

Cornish, R. J. (1974) A graphical analysis for flood control by temporary storage. *J. Instn. Munic. Engrs.*, **101,** 124–7.

Cunge, J. A. (1969) On the subject of a flood propagation computation method (Muskingum method). *J. Hydraul. Res.*, **7,** 205–30.

Dooge, J. C. I. (1973) Linear theory of hydrologic systems. US Dept. Agric., Agric. Res. Serv., Tech. Bull. 1468, 327 pp.

Ellis, J. (1970) Unsteady flow in channel of variable cross-section. *Proc. Am. Soc. Civ. Engrs., J. Hydraul. Div.*, **96**(HY10), 1927–45.

Fletcher, A. G. and Hamilton, W. S. (1967) Flood routing in an irregular channel. *Proc. Am. Soc. Civ. Engrs., J. Engng. Mech. Div.*, **93**(EM3), 45–62.

Gill, M. A. (1978) Flood routing by the Muskingum method. *J. Hydrol.*, **36,** 353–63.

Hall, M. J. and Hockin, D. L. (1980) Guide to the design of storage ponds for flood control in partly urbanised catchment areas. Construction Industry Research and Information Association, Tech. Note 100, 103 pp.

Halliwell, A. R. and Ahmed, M. (1973) Flood-routing in non-prismatic channels using an implicit method of solution. In *Proc. Int. Symp. on River Mech.* (International Association of Hydraulic Research and Asian Institute of Technology, Bangkok) Vol. 3, pp. 263–74.

HARRIS, G. (1970) Real time routing of flood hydrographs in storm sewers. *Proc. Am. Soc. Civ. Engrs., J. Hydraul. Div.*, **96**(HY6), 1247–60.

HAYAMI, S. (1951) On the propagation of flood waves. Kyoto Univ., Disaster Prevention Res. Inst., Bull. No. 1, 16 pp.

HENDERSON, F. M. and WOODING, R. A. (1964) Overland flow and ground-water flow from a steady rainfall of finite duration. *J. Geophys. Res.*, **69,** 1531–40.

JOLLY, J. P. and YEVJEVICH, V. (1974) Simulation accuracies of gradually varied flow. *Proc. Am. Soc. Civ. Engrs., J. Hydraul. Div.*, **100**(HY7), 1011–30.

JONES, S. B. (1981) Choice of space and time steps in the Muskingum–Cunge flood routing method. *Proc. Instn. Civ. Engrs., Part 2*, **71,** 759–72.

KIBLER, D. F. and WOOLHISER, D. A. (1970) The kinematic cascade as a hydrologic model. Colorado State Univ., Hydrol. Paper No. 39, 27 pp.

KOUSSIS, A. (1978) Theoretical estimation of flood routing parameters. *Proc. Am. Soc. Civ. Engrs., J. Hydraul. Div.*, **104**(HY1), 109–15.

LIGHTHILL, M. J. and WHITHAM, G. B. (1955) On kinematic waves: I. Flood movement in long rivers. *Proc. Roy. Soc. Ser. A*, **229,** 281–316.

MITCHELL, T. B. (1962) Lag curves for flood routing through a reservoir. *Proc. Instn. Civ. Engrs.*, **22,** 309–16.

MOZAYENY, B. and SONG, C. S. (1969) Propagation of flood waves in open channels. *Proc. Am. Soc. Civ. Engrs., J. Hydraul. Div.*, **95**(HY3), 877–92.

NATURAL ENVIRONMENT RESEARCH COUNCIL (1975) *Flood Studies Report*, Vol. III: *Flood routing studies* (NERC, London) 76 pp.

PINKAYAN, S. (1972) Routing storm water through a drainage system. *Proc. Am. Soc. Civ. Engrs., J. Hydraul. Div.*, **98**(HY1), 123–35.

PONCE, V. M. (1979) Simplified Muskingum routing equation. *Proc. Am. Soc. Civ. Engrs., J. Hydraul. Div.*, **105**(HY1), 85–91.

PONCE, V. M. and YEVJEVICH, V. (1978) Muskingum–Cunge method with variable parameters. *Proc. Am. Soc. Civ. Engrs., J. Hydraul. Div.*, **104**(HY12), 1663–7.

PRICE, R. K. (1973) Flood routing methods for British rivers. *Proc. Instn. Civ. Engrs., Part 2*, **55,** 913–30.

PRICE, R. K. (1978) A river catchment flood model. *Proc. Instn. Civ. Engrs., Part 2*, **65,** 655–68.

PRICE, R. K. and SAMUELS, P. G. (1980) A computational hydraulic model for rivers. *Proc. Instn. Civ. Engrs., Part 2*, **69,** 87–96.

SARGINSON, E. J. (1969) Streamflow routing analysis. *Civ. Engng. Pub. Wks. Rev.*, **64,** 782–3.

SARGINSON, E. J. (1973) Flood control in reservoirs and storage pounds. *J. Hydrol.*, **19,** 351–9.

SHERMAN, B. and SINGH, V. P. (1976a) A distributed converging overland flow model. 1. Mathematical solutions. *Wat. Resour. Res.*, **12,** 889–96.

SHERMAN, B. and SINGH, V. P. (1976b) A distributed converging overland flow model. 2. Effect of infiltration. *Wat. Resour. Res.*, **12,** 897–901.

SINGH, V. P. (1975) Hybrid formulation of kinematic wave models of watershed runoff. *J. Hydrol.*, **27,** 33–50.

SINGH, V. P. (1976) A distributed converging overland flow model. 3. Application to natural watersheds. *Wat. Resour. Res.*, **12,** 902–8.

SINGH, V. P. and WOOLHISER, D. A. (1976) A non-linear kinematic wave model for watershed surface runoff. *J. Hydrol.*, **31,** 221–43.

SIVALOGANATHAN, K. (1978a) Free surface flow computations by characteristics. *Proc. Am. Soc. Civ. Engrs., J. Hydraul. Div.*, **104**(HY4), 543–56.

SIVALOGANATHAN, K. (1978b) Flood routing by characteristic methods. *Proc. Am. Soc. Civ. Engrs., J. Hydraul. Div.*, **104**(HY7), 1075–91.

SIVALOGANATHAN, K. (1980) Accuracy of explicit methods of unsteady flow computation in channels. *Proc. Instn. Civ. Engrs., Part 2,* **69,** 199–207.

SMITH, A. A. (1980) A generalised approach to kinematic flood routing. *J. Hydrol.*, **45,** 71–89.

SORENSEN, K. E. (1952) Graphical solution of hydraulic problems. *Proc. Am. Soc. Civ. Engrs.*, **78,** Separate No. 116, 17 pp.

STEPHENSON, D. (1979) Direct optimisation of Muskingum routing coefficients. *J. Hydrol.*, **41,** 161–5.

THOMAS, I. E. and WORMLEATON, P. R. (1970) Flood routing using a convective diffusion model. *Civ. Engng. Pub. Wks. Rev.*, **65,** 257–9.

WEINMANN, P. E. and LAURENSON, E. M. (1979) Approximate flood routing methods: a review. *Proc. Am. Soc. Civ. Engrs., J. Hydraul. Div.*, **105**(HY12), 1521–36.

WEST, M. J. H. (1974) Flood control in reservoirs and storage pounds—a discussion. *J. Hydrol.*, **23,** 63–71.

WOODING, R. A. (1965a) A hydraulic model for the catchment-stream problem. I. Kinematic wave theory. *J. Hydrol.*, **3,** 254–67.

WOODING, R. A. (1965b) A hydraulic model for the catchment-stream problem. II. Numerical solutions. *J. Hydrol.*, **3,** 268–82.

WOODING, R. A. (1966) A hydraulic model for the catchment-stream problem. III. Comparison with runoff observations. *J. Hydrol.*, **4,** 21–37.

WOOLHISER, D. A. (1969) Overland flow on a converging surface. *Trans. Am. Soc. Agric. Engrs.*, **12,** 460–2.

WYCOFF, R. L. and SINGH, U. P. (1976) Preliminary design of small flood detention reservoirs. *Wat. Resour. Bull.*, **12,** 337–49.

WYLIE, E. B. (1970) Unsteady free surface flow computations. *Proc. Am. Soc. Civ. Engrs., J. Hydraul. Div.*, **96**(HY11), 2241–51.

YEVJEVICH, V. M. (1964) Bibliography and discussion of flood routing methods and unsteady flow in channels. US Geol. Survey, Water-Supply Paper 1690, 235 pp.

YEVJEVICH, V. and BARNES, A. H. (1970) Flood routing through storm drains, Part IV: Numerical computer methods of solution. Colorado State Univ., Hydrol. Paper No. 46, 47 pp.

# Part III

# HYDROLOGICAL PROBLEMS OF URBAN AREAS

# 8

# The Hydrological Consequences of Urbanisation

## 8.1. INTRODUCTION

The principal function of urban sewerage systems is the elimination of surface water runoff from the built-up area as quickly and as efficiently as possible. Separate stormwater sewers are normally constructed to discharge into a local watercourse at its nearest point. Combined sewer systems generally contain overflow structures that limit the volume of runoff which is carried forward for treatment at the local sewage works. Since stormwater overflows also discharge into the nearest watercourse, the effect of a combined sewerage system on the flow regime of a catchment area is virtually the same as that of a separate system with regard to the quantity of runoff. The increased runoff volume, the shorter time interval before the arrival of the peak flow rate, and the increase in frequency of occurrence of peak discharges of a given magnitude caused by the increase in impervious area and the construction of sewers, all serve to create downstream flooding problems external to the urban area. In many cases, these problems are aggravated by the tendency for urban development to encroach on the flood plains of the local watercourses, thereby reducing the amount of overbank storage.

Until recent years, the solution to these downstream flooding problems has been sought through structural measures, such as the design of flood alleviation channels, flood embankments or flood storage ponds. A variety of non-structural measures, which include land use controls and flood insurance schemes, are now being employed to an increasing extent, particularly in the United States, in an effort to avoid the heavy capital expenditures associated with major urban flood protection works.

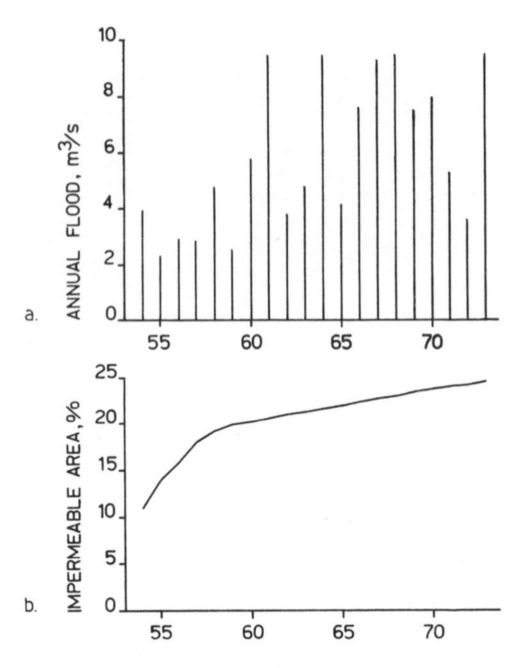


FIG. 8.1. Annual floods and percentage impervious areas for Crawters Brook at Hazelwick Roundabout, 1954–1973.

However, from the hydrological viewpoint, the implementation of such non-structural measures is as demanding in its need for estimates of floods corresponding to different return periods as the design of the engineering works whose construction they are intended to avoid. Indeed, for certain non-structural approaches such as flood insurance programmes, which depend upon a knowledge of the extent of flooding associated with low-frequency events, the scope of the hydrological and hydraulic analyses may exceed that associated with, say, a straightforward channel enlargement.

The development of flood estimation methods for urbanising catchments has been greatly impeded until recent years by a dearth of streamflow and rainfall records relating to such areas. Given that such records are available, the changes to the catchment itself require quantification using Ordnance Survey maps or aerial photographs of different dates, or even satellite imagery (Jackson *et al.*, 1977). Even when both the changes in land use and the alterations to the flow regime have been characterised, their relationship may not be immediately obvious. For example, Fig. 8.1(a) shows the annual maximum instantaneous peak discharges for the Crawters Brook at Hazelwick Roundabout, a catchment area of 4·7 $km^2$ in south-east England which contains part of the new town of Crawley. The increase in the extent of impervious area during the period of record associated with progressive urbanisation is also illustrated in Fig. 8.1(b). The clustering of the largest annual floods during the latter half of the record when the impervious area exceeded 20% of the catchment is obvious. However, this superficial connection ignores climatic variability, which could account for different magnitudes and frequencies of storm events during the period of record, and the influence of changes within the catchment area other than an increase in the extent of paved surfaces. A prerequisite to any study on an urbanising catchment is therefore some means of establishing the significance of the changes in flow regime as indicated by the available observations. The latter is considered in Section 8.2 prior to a review of the available methods for estimating flood peaks and deriving design flood hydrographs for urbanising catchment areas contained in Sections 8.3 and 8.4 respectively.

## 8.2. DETECTION OF CHANGES IN FLOW REGIME

Although the physical changes within a catchment area brought about by urbanisation are clearly obvious, the alterations to the flow regime are more subtle. The increases in runoff volume and the lower return periods of flows of a given magnitude manifest themselves in changes in the statistics that characterise the frequency distributions which describe different features of the streamflow record. The extent to which such changes take place may be evaluated either by the application of statistical tests of significance or by the use of double-mass curve analysis. In most cases the annual runoff volumes for the catchment area of interest are generally found to be sufficient to indicate the relative magnitude of any changes and their direction.

In order to apply the statistical tests of significance, the series of annual runoff volumes is divided into two non-overlapping sub-sets of size $n_1$ (pre-development) and $n_2$ (post-development), and their means, $m_1$ and $m_2$, and standard deviations, $s_1$ and $s_2$, are computed. Firstly, the variance ratio or $F$-test is applied to the variances. The test statistic

$$F = s_1^2/s_2^2; \qquad s_1^2 > s_2^2 \tag{8.1}$$

is computed and compared with the tabulated value of $F$ for $n_1-1$ and $n_2-1$ degrees of freedom and a confidence level of $\alpha/2$. A value of $\alpha$ of 5% is adequate for the majority of applications, and tables of the statistic may be found in most statistical textbooks and manuals, such as Crow *et al.* (1960). If the computed value exceeds the tabulated value, the null hypothesis that $s_1^2$ and $s_2^2$ are estimates of the same population variance is rejected. However, if the computed value is less than the tabulated value, the null hypothesis is accepted and a pooled estimate of variance may be formed from

$$s^2 = [(n_1-1)s_1^2 + (n_2-1)s_2^2]/(n_1+n_2-2) \tag{8.2}$$

This pooled estimate is required for the second test of significance, referred to as the $t$-test, which is applied to the means of the two sub-sets. The test statistic

$$t = |(m_1 - m_2)|/s\sqrt{(1/n_1 + 1/n_2)} \tag{8.3}$$

is computed and compared with the tabulated value of $t$ for $n_1+n_2-2$ degrees of freedom and a confidence level of $\alpha/2$. If the computed value exceeds the tabulated value, the null hypothesis that $m_1$ and $m_2$ are estimates of the same population mean is rejected.

The application of the above-mentioned tests may be conveniently illustrated using the annual flows for the catchment area of East Meadow Brook on Long Island, New York, tabulated by Sawyer (1963). This catchment was subjected to intensive urban development from 1952 onwards. The available flow records could therefore be divided into two sub-sets consisting of 14 years (1938–1951) pre-development and 9 years (1952–1960) post-development. The means and standard deviations for these periods may be summarised as follows:

| *Period* | *Number of data* | *Mean flow (mm)* | *SD (mm)* |
|---|---|---|---|
| 1938–1951 | 14 | 185 | 33 |
| 1952–1960 | 9 | 214 | 30 |

The variance ratio test gives a computed $F$-statistic of 1·21, which compares with a tabulated value of 4·18 for 13 and 8 degrees of freedom and a significant level of 2·5%. The null hypothesis that the variances are estimates of the same population variance can therefore be accepted. Using eqn (8.2), the pooled estimate of variance is found to be 1017 $mm^2$. The computed $t$-statistic is therefore 2·13. Since the tabulated $t$-statistic for 21 degrees of freedom and a significant level of 2·5% is 2·09, this second test is sufficient to cast doubt on the null hypothesis that the means of the sub-sets are estimates of the same population mean; more data would be required and the tests repeated before a definite conclusion could be drawn.

Double-mass curve analysis may also be used in addition to statistical significance tests as a means of evaluating a time series of flows for changes in their character. Such curves consist of accumulated values of a test variable plotted against accumulated values of a reference variable for a concurrent period. If the data are proportional, the graph will plot as a straight line, but a break in slope indicates either a change in the proportionality of the variables or that the proportionality is not constant for all rates of accumulation. Ignoring the latter possibility, the break indicates the time at which the change took place.

One of the principal difficulties that can arise in employing observed hydrological time series as both test and reference variables is the problem of identifying which of the records has caused the break in slope. Indeterminate results can be avoided by using the average of several concurrent data sets as the reference variable, an approach which has found widespread use in testing the consistency of rainfall observations (see, for example, Kohler, 1949). Even then, double-mass curve analysis depends upon the test and reference variables, $y$ and $x$, being in constant ratio:

$$y = cx; \qquad y = c(x - a); \qquad (y - b) = cx$$

Such relationships are rarely found to be applicable if one data set is streamflow and the other is rainfall, and Searcy and Hardison (1960) have suggested a modified procedure for such cases which may be summarised as follows:

(1) list the test variable (runoff) in date order and rank in descending order of magnitude;
(2) list the reference variable (rainfall) in date order and rank as in step 1;
(3) compute the sums of squares of the differences in concurrent ranks of the two series;
(4) compute an 'effective rainfall' equal to a weighted moving

average of the rainfalls for the current and preceding years, the weights summing to unity, and rank as in step 2;

(5) repeat steps 3 and 4 for different values of the weights until the sum of squares reaches a minimum value;

(6) carry out a linear regression of runoff on the effective rainfall defined in terms of the chosen weights; and

(7) use the estimates of runoff from the regression equation as the reference variable to test the time series of observed runoff.

When interpreting the results of a double-mass curve analysis, the possibility that spurious breaks in slope may sometimes result from the inherent variability of hydrological data should be borne in mind, and any changes that do not persist for more than 5 years should be ignored. Provided that the basic assumption of test and reference variables being in constant ratio is satisfied, such analyses are capable of yielding informative results. The technique has been applied to evaluate the effects of constructing sewerage systems for stormwater and piping wastewater to treatment plants on groundwater levels in a rapidly urbanising area of Long Island, New York, by Franke (1968) and Garber and Sulam (1976). Double-mass curve analysis was also used by Harris and Rantz (1964) to examine the changes in flow regime brought about by urban development within the catchment of an ephemeral stream in California.

Where large numbers of data sets are to be processed, double-mass curve analysis can be tedious to perform without the aid of a digital computer. Suitable algorithms, that can be applied with minimal intervention from the analyst, have been described by Singh (1968) and Chang and Lee (1974).

## 8.3. CHANGES IN THE FREQUENCY DISTRIBUTION OF FLOODS

The fitting of frequency distributions to samples of observed flood peaks has already been discussed in Chapter 5. Whichever of the standard statistical distributions is chosen, the recorded peak flows are assumed to form a homogeneous data set and to be unaffected by external influences, a condition that is obviously invalid for a catchment area undergoing urban development. Unless urbanisation takes place within a relatively short period, such that the pre-development and post-development records can be treated as two separate data sets, some method is required for reducing the observed peak flows to a standard catchment state. Several authors, including James (1965), Gundlach (1978) and Beard and Chang (1979), have advocated deterministic modelling as a method of solving this problem.

In this approach, records of both rainfall and streamflow, obtained from a catchment whose state of development has remained unchanged, are used to calibrate a rainfall–runoff model. Once a satisfactory agreement has been obtained between observed and synthesised hydrographs, the model parameters are altered to reflect the influence of urban development. The model may then be employed to generate a series of synthetic hydrographs that are representative of different stages of urban development from the historical rainfall record. An analysis of the annual flood series provided by each synthetic time series will yield the frequency distributions from which the changes in magnitude of floods of different return period may be assessed.

This procedure was applied by James (1965) to construct a series of diagrams in which the magnitude of the flood of a given return period was related to the percentage of the catchment which had been urbanised and the percentage of the channels that had been improved. These calculations related specifically to a 113 $km^2$ drainage basin in California, and were carried out using a variant of the Stanford Watershed model (see Crawford and Linsley, 1966). Although the model employed by both Gundlach (1978) and Beard and Chang (1979) was much simpler in form and based upon the Unit Hydrograph Method, its application is subject to similar criticisms. Firstly, a convincing agreement must be obtained between observed and synthesised streamflows during the calibration of the model. Secondly, the model parameters which are altered to simulate the influence of urban development must possess an appropriate physical interpretation rather than simply a mathematical significance as coefficients in a curve-fitting exercise. With the more complex models, such as that employed by James (1965), interaction between model parameters may compound the difficulties of applying a physical interpretation. For the simpler models, the relationships between model parameters and catchment characteristics may not be capable of sufficient refinement to capture the subtleties of the changes in flow regime. Much work remains to be done in proving the validity of this deterministic modelling approach.

An alternative method of evaluating the changes in the form of a flood frequency distribution caused by urbanisation, which has received considerable attention, attempts to correct the observed flood peaks within the framework of a regional analysis. In this approach, the effects of urban development on the magnitude of an index flood, such as the mean annual flood, and the growth factors, expressing the ratios of $T$-year floods to the index flood, are treated independently.

Carter (1961), who was among the first to investigate the changes in the frequency distribution of floods brought about by urbanisation, made the basic assumption that the influences of increasing impervious area and extensions to sewerage systems could be allowed for separately. The effect of impervious area was characterised by means of a coefficient of imperviousness, CIM, based upon the premise that 30% runoff could be expected from rural catchments and 75% from completely impervious basins:

$$\begin{aligned} \mathrm{CIM} &= [0{\cdot}3+(0{\cdot}75-0{\cdot}3)\,\mathrm{IMP}/100]/0{\cdot}3 \\ &= 1+0{\cdot}015\,\mathrm{IMP} \end{aligned} \tag{8.4}$$

where IMP is the percentage impervious area. The value of CIM therefore ranges from 1·0 for zero impervious area to 2·5 for 100% impervious cover.

The influence on catchment response of sewerage systems was characterised by the lag time, defined as the time interval between the centroid of the rainfall hyetograph and the centroid of the runoff hydrograph. This parameter was related to the basin ratio, i.e. the quotient of the main channel length and the square root of the main channel slope, and separate equations derived for undeveloped (rural), partially sewered and completely sewered (urban) catchment areas. An expression for the mean annual flood of a catchment area, $\bar{Q}$, adjusted for the effects of impervious area by means of CIM, was then obtained in terms of catchment area, A, and lag time, $TL$:

$$\bar{Q}/\mathrm{CIM} = cA^{a}TL^{b} \tag{8.5}$$

where $a$ and $b$ are exponents and $c$ is a constant whose values were obtained by multiple linear regression analysis.

This approach to the estimation of the mean annual flood for urbanising catchments was also adopted by Martens (1968) and Anderson (1970), who both extended their analyses to floods of a higher return period. In each of these studies the growth factors, $R = Q(T)/\bar{Q}$, where $Q(T)$ is the flood having a return period of $T$ years, for undeveloped conditions were obtained from annual flood series from rural drainage basins within the region of interest. For completely impervious catchments the growth factors were assumed to approach those of the rainfall, and were obtained by averaging the ratios of the $T$-year rainfall depths to the mean annual rainfall depths over a range of durations. Given the growth factors for the rural (undeveloped) catchments, $R_n$, and the completely impervious drainage basins, $R_u$, those corresponding to an impervious area of IMP, $R_i$, were interpolated using the equation

$$R_i = [R_n + (2{\cdot}5R_u - R_n)\mathrm{IMP}/100]/\mathrm{CIM} \tag{8.6}$$

This method of quantifying the changes in the magnitude of higher return period floods may be contrasted with that of Espey and Winslow (1974), in which peak flow rates of up to a 50-year return period were used as the dependent variables in a multiple linear regression analysis with catchment characteristics as the independent variables. The latter included both the percentage of impervious area and an empirical factor that varied with both the extent of channel improvements and storm sewer construction, and the amount of channel vegetation.

Although the basis of the interpolated growth factors is largely intuitive, eqn (8.6) reflects the consensus of opinion identified by Hollis (1975) that, in general, the smaller, more frequent floods are affected more by urban development than the larger events. The low return period rainfalls that are merely absorbed by the available soil moisture storage on a rural drainage basin are capable of generating runoff from the impervious surfaces of urban catchments. However, during more severe rainstorms, soil moisture deficits are satisfied more quickly and the channel network of a rural catchment area extends so that its flood response becomes comparable to that of an urban area. This result has particular significance for the planning of flood control measures, since the larger and rarer floods, which have the potential to cause the most damage and disruption, are the least affected by urbanisation.

As noted in Section 4.3, the regional flood frequency analysis for the British Isles presented in the Flood Studies Report (Natural Environment Research Council, 1975) was based upon six-variable equations for six regions and a three-variable equation for the seventh. Only the three-variable equation contained an independent variable, URBAN, defined as the fraction of the catchment area shown as developed on a 1:50 000 Ordnance Survey map, which provided some measure of urbanisation. Methods for modifying the results of the regional analysis to reflect the influence of urban development on both the magnitude of the mean annual flood and the shape of the growth curve were subsequently presented by the Institute of Hydrology (1979). The change in the mean annual flood was found to be described by the equation

$$\bar{Q}_u/\bar{Q}_r = (1+\text{URBAN})^{2NC}[1+\text{URBAN}\{(21/\text{CIND})-0\cdot3\}] \tag{8.7}$$

where the subscripts u and r denote urban and rural conditions respectively; $NC$ is a rainfall continentality factor, equivalent to the exponent $n$ in eqn (3.4); and CIND is a catchment index depending upon a soil index and a catchment wetness index, the latter being a function of average annual rainfall.

Owing to the dearth of long streamflow records from drainage areas in various stages of urban development, the description of the effect of urbanisation on the shape of growth curves was largely intuitive. Analysis of the available data showed that urban and rural growth factors were approximately equal around a frequency of once in 7 years, but $R_u$ exceeded $R_r$ at larger return periods, with $R_u - R_r$ increasing with increasing values of URBAN. For return periods up to 50 years this variation was described by tabulating 'equivalent return periods' such that $R_u$ for the urban $T$-year flood is found at a return period $T'$ on the rural growth curve. For return periods exceeding 50 years the ratio of the urban $T$-year flood, $Q_u(T)$, to the rural $T$-year flood, $Q_r(T)$, was assumed to decay exponentially according to the expression

$$Q_u(T)/Q_r(T) = 1 + B\exp(-ky) \tag{8.8}$$

where $y$ is the extreme value type 1 standardised variate given by eqn (5.59), and $k$ and $B$ are constants. Using the ratios at return periods of 6·6 years (where $Q_u(T)/Q_r(T) = \bar{Q}_u/\bar{Q}_r$) and 50 years, the following expressions are obtained for $k$ and $B$ after some simplification and grouping of terms:

$$k = 0\cdot48\ln[(1-\text{RMAF})/(G(50)-\text{RMAF})] \tag{8.9}$$

$$B = ((1/\text{RMAF})-1)\exp(1\cdot8k) \tag{8.10}$$

where $\text{RMAF} = \bar{Q}_r/\bar{Q}_u$, and $G(50)$ is the ratio $R_u/R_r$ at a return period of 50 years. The ratio of urban to rural growth factors for return period

$T$ then becomes

$$G(T) = \text{RMAF} + (1 - \text{RMAF}) \exp\{-k(\ln T - 1{\cdot}8)\} \qquad (8.11)$$

where the approximation $y = \ln T$ is made in place of eqn (5.59). The urban $T$-year flood follows from eqn (8.11), given the rural growth factor at return period $T$ and the urban mean annual flood from eqn (8.7).

## 8.4. CHANGES IN DESIGN FLOOD HYDROGRAPHS

The discussion in Chapter 4 of flood estimation methods has already drawn attention to the dominance in engineering practice of techniques based upon the unit hydrograph approach as a means of producing a design flood hydrograph. One of the principal assumptions of the Unit Hydrograph Method is time-invariance, i.e. a constancy in the relationship between rainfall and runoff with time. Since a drainage area undergoing urbanisation will have a markedly time-variant rainfall–runoff relationship, the shape of the unit hydrograph can be expected to change as urban development proceeds. If flow records spanning the onset of urbanisation are available for the catchment of interest, an analysis of storm events which are representative of different stages of urban growth may be employed to characterise the changes in flood response. However, in order to be able either to forecast the effects of further growth in a gauged catchment, or to evaluate the flooding potential of an ungauged urban drainage basin, a more generalised approach based upon the available data from areas in different stages of urban development must be followed. This approach involves the derivation of the unit hydrograph for a common rainfall duration for each of the gauged catchments, the selection of certain key dimensions to define their shape, and the regression of these dimensions on catchment characteristics.

One of the most critical of these dimensions is the lag time, for which at least eight definitions, including the time-to-peak, may be identified, as shown in Fig. 8.2. Perhaps the most extensively used of these measures of catchment response is that labelled $T3$, the time interval between the centre of mass of the distribution of rainfall excess and the centre of mass of the direct runoff hydrograph. However, $T2$, $T3$ and $TP$ have all been employed as a scaling parameter in defining a common form of dimensionless unit hydrograph, and have appeared as

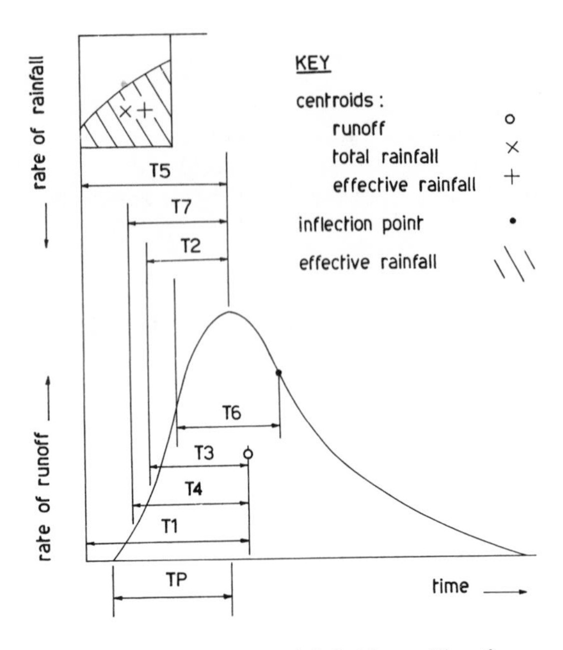


FIG. 8.2. Summary of definitions of lag time.

dependent variables in a regression analysis on catchment characteristics, including measures of urban development. Although this approach has been common to many investigations into the effect of urbanisation on the flood behaviour of catchments, each has tended to differ in detail, particularly with regard to the shape of the dimensionless unit hydrograph.

One of the first of the so-called synthetic unit hydrograph methods, i.e. techniques for defining the unit hydrograph for an ungauged area, was that described by Snyder (1938). Using data from the Appalachian Highlands of the United States, Snyder assumed that, for storms of a given type on a particular catchment, the lag time, $T2$, was constant and defined by the expression

$$T2 = CT(LCA \,.\, L)^{0\cdot 3} \tag{8.12}$$

where $LCA$ is the distance between the gauging station and the centroid of the catchment along the main channel, $L$ is the total length of the catchment, and $CT$ is an empirical coefficient, $1\cdot 8 \leqslant CT \leqslant 2\cdot 2$, with an average value for the Appalachian Highlands of 2·0. The peak discharge per unit area of the unit hydrograph, $QP$, was related to the lag time by a second empirical equation:

$$QP = CP/T2 \tag{8.13}$$

where $CP$ is a second coefficient, $360 \leqslant CP \leqslant 440$, with an average value of 400. (The quoted values of $CT$ and $CP$ apply with $LCA$ and $L$ in miles, $T2$ in hours and $QP$ in $ft^3/s/mile^2$). Finally, the time base of the unit hydrograph in days was related to $T2$ by the expression

$$TB = 3 + 3(T2/24) \tag{8.14}$$

Equations (8.12)–(8.14) applied to a unit hydrograph of duration $\Delta t = T2/5{\cdot}5$. According to Johnstone and Cross (1949), this choice of duration arose from the division of the rising limb of the direct runoff hydrograph into six equal time steps. The time-to-peak was therefore $6\,\Delta t$, which was also equal to $T2 + \Delta t/2$ (see Fig. 8.3(b)); $T2$ was therefore 5·5 times the unit duration. In order to construct the unit

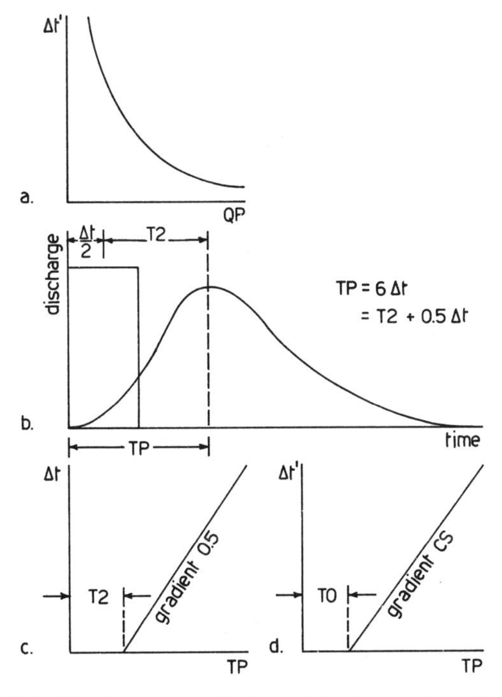


FIG. 8.3. The Snyder synthetic unit hydrograph method.

hydrograph for a different duration of effective rainfall, $\Delta t'$, Snyder (1938) suggested the substitution of an adjusted lag time, $T2'$, where

$$T2' = T2 + 0{\cdot}25(\Delta t' - \Delta t) \tag{8.15}$$

$$QP = CP/T2' \tag{8.16}$$

The choice of the correction term, $0{\cdot}25(\Delta t' - \Delta t)$, was not fully explained in the original paper. Its justification is apparently purely empirical in that eqn (8.16) allows for the reduction in the peak ordinate of the unit hydrograph as its duration increases (see Fig. 8.3(a)).

Linsley (1943), who applied the Snyder method to 18 drainage areas in the Central Valley of California, was critical of the assumption of a constant lag time, the corollary to which is that a plot of the time-to-peak against rainfall duration for selected storm events should be a straight line having a gradient of one-half and an intercept equal to the lag time (see Figs 8.3(b),(c)). When such diagrams were constructed for the Californian catchments studied by Linsley, straight-line relationships were obtained whose gradient, $CS$, exceeded $0{\cdot}5$, and were better represented by the equation

$$TP = T0 + CS\,\Delta t' \tag{8.17}$$

where $T0$ is the intercept on the $TP$-axis (see Fig. 8.3(d)), and $0{\cdot}7 \leqslant CS \leqslant 1{\cdot}0$, with an average value of $0{\cdot}85$. Substituting for $TP = T2 + \Delta t'/2$ in eqn (8.17), and replacing $T2$ by an adjusted lag time, $T2''$,

$$T2'' = T0 + (CS - 0{\cdot}5)\,\Delta t' \tag{8.18}$$

Equation (8.18) is similar in form to eqn (8.15). Using the average value of $CS$, the correction term of eqn (8.18) becomes $0{\cdot}35\,\Delta t'$; if $\Delta t$ is assumed to be small, $T2$ can be approximated by $T0$ and the correction term of eqn (8.15) becomes $0{\cdot}25\,\Delta t'$. These similar features led Linsley to suggest that eqn (8.13) should be employed to compute the peak discharges for unit hydrographs corresponding to any rainfall duration. For the Californian catchments, $225 \leqslant CP \leqslant 320$, with an average value of 270 (Imperial units). However, the lag time of eqn (8.12) was constrained to apply to a standard rainfall duration, and $T0$, which was considered applicable to a very short storm, was adopted. Using this variable, $0{\cdot}3 \leqslant CT \leqslant 0{\cdot}7$, with an average value of $0{\cdot}5$ (Imperial units).

Despite the expected variation in the values of $CP$ and $CT$, the study by Linsley (1943) provided useful corroboration for the general form of eqns (8.12)–(8.15). As with all such empirical equations, their extrapolation to both other regions and other types and sizes of catchment area must be carried out with caution. Snyder (1938) considered his approach applicable to drainage areas between 26 and 26 000 $km^2$, and the catchments studied by Linsley (1943) ranged from 260 to 9600 $km^2$, all of which were essentially rural drainage basins. The question therefore arises as to whether the same relationships hold for both urbanising and fully developed urban catchments and, if so, whether appropriate values of *CT* and *CP* can be found to characterise their behaviour.

Eagleson (1962) was among the first to apply the Snyder method to fully sewered urban areas. The 5 catchments included in his study, ranging in size from 0·57 to 19·45 $km^2$, were much smaller than those investigated by Snyder and Linsley, making the use of eqn (8.14) for *TB* completely inappropriate. However, eqns (8.12) and (8.13) were found to be applicable, with much smaller values of *CT* and *CP* towards the lower end of the range quoted by Snyder (1938). Nelson (1970), who analysed the 1 h unit hydrograph for 8 catchments in the Fort Worth–Dallas area including completely rural and fully urban areas, also found lower *CT* values, but quoted a *CP* of 450 (Imperial units) in tolerable agreement with Snyder (1938). This tendency for *CT* to vary with the extent of urban development, but for *CP* to remain relatively constant, was also noted by Bleek (1975) in a study of urbanising areas in south-east England. The latter author found *CT* to be inversely proportional to the square root of the percentage of impervious area. In contrast, Van Sickle (1969) has maintained that the ratio of *CP* to *CT* is a better descriptor of urban development. The generality of either relationship has yet to be confirmed.

Once appropriate values of *T*2 and *QP* have been estimated for a drainage area, the position of the peak of the unit hydrograph is fixed, and its shape must be sketched subject to the condition that a unit volume of (say) 10 mm of rainfall excess over the catchment must be contained under the curve. This operation is greatly assisted if other dimensions of the unit hydrograph, such as the time base or the widths of the hydrograph at different proportions of the peak flow rate, can also be predicted, as demonstrated by Espey *et al.* (1965).

The principal determinant of unit hydrograph shape employed by Espey *et al.* (1965) was the time-to-peak. Using data from 11 rural catchments ranging from 0·35 to 18·2 $km^2$ and 22 urban drainage basins betwen 0·03 and 238·3 $km^2$ in an area located mainly in the eastern states of America, those authors derived separate equations for the time-to-peak of the 30 min unit hydrograph. For the rural areas,

$$TP_r = 2{\cdot}65L^{0{\cdot}12}S^{-0{\cdot}52} \tag{8.19}$$

where $L$ is the main channel length (ft), $S$ is the main channel slope (dimensionless) and $TP_r$ is measured in minutes. Equation (8.19) was found to explain 95% of the variance of $TP_r$. For the urban areas,

$$TP_u = 20{\cdot}8L^{0{\cdot}29}S^{-0{\cdot}11}\text{IMP}^{-0{\cdot}61} \tag{8.20}$$

where IMP is the percentage impervious area. Although eqn (8.20) explained 91% of the variance in $TP_u$, predicted values were found to overestimate measured values on catchments that had had either extensive channel improvements carried out or stormwater sewers installed. This performance was attributed to the inadequacy of IMP as an index of urbanisation, and so an empirical factor, PHI, was therefore introduced to account for the observed reductions in time-to-peak. After some rounding, the following classification was adopted for PHI:

| *PHI* | *Classification* |
|---|---|
| 0·6 | Extensive channel improvement and storm sewer system; closed conduit channel system |
| 0·8 | Some channel improvement and storm sewers; mainly cleaning and enlargement of existing channel |
| 1·0 | Natural channel conditions; no urban development |

Equation (8.20) was therefore written

$$TP_u = 20{\cdot}8\text{PHI} \,.\, L^{0{\cdot}29}S^{-0{\cdot}11}\text{IMP}^{-0{\cdot}61} \tag{8.21}$$

Further examination of the available data showed that, for both urban and rural catchments, the peak discharge of the 30 min unit hydrograph was a function of both the catchment area and the time-to-peak. The data were therefore combined to derive a composite equation for the peak discharge per unit area, $QP$:

$$QP = 40\,900\,TP^{-1{\cdot}11} \tag{8.22}$$

where $QP$ is measured in $ft^3/s/mile^2$. Equation (8.22) explained 90% of the variance in $QP$.

An analysis of hydrograph widths also showed that both catchment area and peak discharge per unit area consistently appeared as the independent variables. Again, composite equations were derived by combining the data sets to give

$$TB = 318\,000 QP^{-1\cdot13} \tag{8.23}$$

$$W50 = 38\,800 QP^{-1\cdot025} \tag{8.24}$$

$$W75 = 10\,000 QP^{-0\cdot89} \tag{8.25}$$

where *TB*, *W*50 and *W*75 are the time base, and the hydrograph widths at 50 and 75% of the peak discharge (min). Equations (8.23)–(8.25) were found to explain 95, 95 and 94% of the variance in the dependent variable respectively.

In a subsequent study, Espey *et al.* (1969) expanded their data set to 17 rural and 33 urban catchments using data obtained from gauging stations in the vicinity of Houston, Texas. This analysis resulted in the revision of the exponents and constants to several of the original equations. In addition, the seasonal variations in channel vegetation were identified as an important influence on $TP_u$, and the empirical factor, PHI, was redefined as the sum of two coefficients, PHI1 and PHI2, where PHI1 was the original PHI and PHI2 was classified as follows:

| *PHI*2 | *Classification* | *PHI*2 | *Classification* |
|---|---|---|---|
| 0 | No vegetation | 0·2 | Moderate vegetation |
| 0·1 | Light vegetation | 0·3 | Heavy vegetation |

Using this revised definition, eqn (8.21) became

$$TP_u = 16{\cdot}4\mathrm{PHI}\,.\,L^{0\cdot32} S^{-0\cdot049} \mathrm{IMP}^{-0\cdot49} \tag{8.26}$$

and eqn (8.19) became

$$TP_r = 2{\cdot}68 L^{0\cdot22} S^{-0\cdot30} \tag{8.27}$$

In order to predict the peak discharge for the 30 min unit hydrograph, the data sets were considered separately. For the urban areas,

$$QP = 35\,000 TP_u^{-1\cdot10} \tag{8.28}$$

but for the rural drainage basins, catchment area, *A* (mile$^2$), was employed as an independent variable, leading to the equation

$$Q_p = 82\,500 A^{0\cdot99} TP_r^{-1\cdot25} \tag{8.29}$$

where $Q_p$ is the peak discharge in ft$^3$/s.

The problems associated with sketching in a hydrograph shape that has the appropriate dimensions obtained from the regression equations and satisfies the constraint of unit volume may be avoided by the use of a standard form defined either by an equation or a simple geometrical approximation. For example, the two-parameter gamma function, which was used by both Edson (1951) and Gray (1961) to describe the geometry of synthetic unit hydrographs for rural drainage areas, has recently been employed for catchments subjected to urban development by Cruise and Contractor (1980). In this approach, the unit hydrograph ordinate,

$$q = [Cy(yt)^x \exp(-yt)]/\Gamma(x+1) \tag{8.30}$$

where $C$ is a conversion constant and $\Gamma$ is the gamma function. According to Edson (1951), the parameter $x$ depends upon the shape of the time–area diagram of the catchment, and the parameter $y$ is a recession constant. At the peak of the unit hydrograph defined by eqn (8.30),

$$t = TP = x/y \tag{8.31}$$

and

$$q = QP = [Cy(x/e)^x]/\Gamma(x+1) \tag{8.32}$$

where $e$ is the base of Napierian logarithms. Since $x$ and $y$ define $QP$ and $TP$ completely, they may be employed as the dependent variables in a regression analysis on catchment characteristics. Using data from 30 catchments in North Carolina and northern Virginia, and basin ratios and percentage impervious areas as the independent variables, Cruise and Contractor (1980) found that this regression model had no generality. The data set had to be divided according to geographical location before significant relationships could be obtained. $x$ was found to be linearly related to the logarithm of the basin ratio and the percentage impervious area, but $y$ depended only on the former.

The two-parameter gamma distribution also describes the form of the instantaneous unit hydrograph (IUH) for a conceptual model consisting of a cascade of $n$ linear reservoirs, each having the same storage constant, $K$. If the parameters, $x$ and $y$, are transformed according to

$$x = n+1; \qquad y = 1/K$$

eqn (8.25) becomes identical to eqn (6.42), as derived by Nash (1957). The application of this conceptual model and the simpler single linear reservoir to urbanising catchments has been investigated by Rao *et al.*

(1972). Those authors found that the model parameters varied from storm to storm at the same site. The rainfall–runoff relationship was therefore assumed to be linear for individual storm events, but the model parameters were allowed to vary between storms by including the volume and duration of effective rainfall as independent variables in the prediction equations. For areas of less than 13 $km^2$, a single linear reservoir model was found to provide an adequate description of catchment behaviour. The storage constant of this reservoir, which is also equal to the lag time, $T3$ (see Fig. 8.2), was obtained from the equation

$$T3 = 0{\cdot}831 A^{0{\cdot}458}(I+1)^{-1{\cdot}66} P_e^{-0{\cdot}267} D^{0{\cdot}371} \tag{8.33}$$

where $T3$ is measured in hours, $A$ is the catchment area ($mile^2$), $I$ is the proportion of impervious area, and $P_e$ and $D$ are the volume (in) and duration (h) of the effective rainfall respectively. Based on 125 storms from 11 drainage areas, eqn (8.33) explained 85·1% of the variance in $T3$.

For larger catchments of between 13 and 52 $km^2$, the cascade of linear reservoirs was found to be the better model. Since for this model $T3 = nK$, the parameters could be found by using eqn (8.33) for $T3$ and a second equation for $K$:

$$K = 0{\cdot}575 A^{0{\cdot}389}(I+1)^{-0{\cdot}622} P_e^{-0{\cdot}106} D^{0{\cdot}222} \tag{8.34}$$

Equation (8.34) explained 72·5% of the variance of $K$, which was again measured in hours. In both eqns (8.33) and (8.34), the impervious area index, $I+1$, is the dominant independent variable.

As an alternative to the use of a linear conceptual model having an analytical form of instantaneous unit hydrograph, the assumption can be made that unit hydrographs of the same duration from a group of urbanising catchment areas are reducible to a common dimensionless form. This approach was applied by Hall (1974, 1977) to 8 catchments in West Sussex and North London. In both studies, the 1 h unit hydrographs representing a particular state of development on individual catchments were made dimensionless by dividing their abscissae and multiplying their ordinates by the centroid-to-centroid lag time, $T3$. The functional form of each dimensionless unit hydrograph was therefore given by

$$u_t T3 = f(t/T3) \tag{8.35}$$

where $u_t$ is the ordinate of the unit hydrograph (reciprocal hours) at time $t$. In general, these dimensionless unit hydrographs were found to reach a peak of 0·8 units at between $0{\cdot}6T3$ and $0{\cdot}8T3$ and to have a time base of approximately $3{\cdot}5T3$.

Using the combined data set of 187 events, Hall (1981) obtained representative responses for specified degrees of urbanisation on each catchment by fitting a polynomial function to each set of dimensionless unit hydrographs by the method of least squares. Denoting the product $u_tT3$ by $Y$ and the quotient $t/T3$ by $X$, the fitted function was given by the expression

$$Y = \sum_{j=0}^{m} C_j X^j \tag{8.36}$$

where the $C_j$, $j = 0, 1, \ldots, m$, are the coefficients of the polynomial of the order $m$. The same approach was applied to derive a general dimensionless 1 h unit hydrograph for the whole data set, and an 8th-order polynomial was found to afford the best compromise between the number of coefficients, the root mean square residual and the area enclosed up to an abscissa of $3{\cdot}5T3$.

Owing to the small number of catchment areas in the data set, suitable relationships between the scaling parameter, $T3$, and catchment characteristics were constructed on the basis of trends exhibited in previously published studies, notably Carter (1961) and Anderson (1970). Following those authors, lag time was plotted against the basin ratio and separate relationships deduced for both rural and urban catchments in south-east England:

$$T3_\mathrm{r} = 0{\cdot}867R^{0{\cdot}42} \tag{8.37}$$

$$T3_\mathrm{u} = 0{\cdot}212\sqrt{R} \tag{8.38}$$

where $R = 31{\cdot}6L/\sqrt{S}$ is the basin ratio, with $L$ the main channel length (km) and $S$ the main channel slope (m/km). The latter was defined as the altitude difference between two points located 10 and 85% of the main channel length upstream from the gauging station divided by $0{\cdot}75L$. The subscripts r and u denote rural and urban conditions respectively, the latter corresponding to a condition with about 25% impervious area and some sewering and channel improvements. For intermediate states of development with a percentage impervious area IMP, the lag time, $T3_\mathrm{i}$, was estimated by logarithmic interpolation between $T3_\mathrm{r}$ and $T3_\mathrm{u}$:

$$\ln T3_\mathrm{i} = \ln T3_\mathrm{r} + 0{\cdot}04\mathrm{IMP} \ln (T3_\mathrm{u}/T3_\mathrm{r}) \tag{8.39}$$

The representation of the shape of the dimensionless unit hydrograph by means of an 8th-order polynomial presents little difficulty when either a digital computer or a programmable calculator is used to construct the design storm hydrograph. For manual calculations, a list of the ordinates is more convenient. Computations may be further simplified by the use of a geometrical approximation to the shape of the dimensionless unit hydrograph instead of a functional form, such as eqns (8.30) or (8.36). Such an approximation was employed as part of the design flood hydrograph method contained in the United Kingdom Flood Studies Report (Natural Environment Research Council, 1975). In this approach, the dimensionless 1 h unit hydrograph was represented by straight line rising and recession limbs, and was completely defined in terms of the time-to-peak, *TP* (h), by the equations

$$TP = 46{\cdot}6 S^{-0{\cdot}48} RSMD^{-0{\cdot}42} L^{0{\cdot}14} (1 + \mathrm{URBAN})^{-1{\cdot}99} \qquad (8.40)$$

$$QP \,.\, TP = 220 \qquad (8.41)$$

$$TB = 2{\cdot}525\, TP \qquad (8.42)$$

where $L$ is the length (km) and $S$ is the slope (m/km) of the main channel; *RSMD* is the 5-year, 1-day rainfall excess (mm), which is a function of the average annual rainfall; URBAN is the fraction of urbanised area in the catchment; *QP* is the peak discharge of the unit hydrograph ($m^3/s/10^2\ km^2$); and *TB* is the time base (h). Equations (8.40)–(8.42) have also been applied to synthesise the unit response of urbanising catchment areas by the Institute of Hydrology (1979).

The above examples are sufficient to demonstrate that the majority of synthetic unit hydrographs that have been proposed for drainage areas subject to urbanisation have employed some measure of lag time as the only scaling parameter. This approach depends primarily on the ability to predict lag time from readily calculable catchment characteristics, a technique which has been found viable at least within homogeneous geographical regions. However, the unit hydrograph is only one component in a design flood estimation method. As discussed in Chapter 4, the choice of the design variables required to maintain a predetermined relationship between the return period of the design storm and the return period of the peak flow rate is equally, if not more, important. A simulation exercise of the type required to identify such design variables has been carried out for two methods. The first is that set out in Flood Studies Supplementary Report No. 5 (FSSR5) by the Institute of Hydrology (1979); and the second is that contained in a

**Table 8.1. Comparison between the principal features of the design flood estimation methods for urbanising catchments proposed in Flood Studies Supplementary Report No. 5 (Institute of Hydrology, 1979) the CIRIA Guide (Hall and Hockin, 1980) and the Flood Studies Report method for rural areas (NERC, 1975)**

| *Component of method* | *Flood Studies Report* | *Flood Studies Suppl. Rept. 5* | *CIRIA Guide* |
|---|---|---|---|
| Design storm profile | 75% winter profile | ← 50% summer profile → | |
| Design storm return period | Non-linear function of return period of peak flow rate | ← Equal to return period of peak flow rate → | |
| Design storm duration | Function of scaling parameter for dimensionless unit hydrograph and average annual rainfall | | |
| Dimensionless unit hydrograph | ← Triangular approximation → | | Polynomial function |
| Scaling parameter for dimensionless unit hydrograph | Time-to-peak, a function of main stream length and slope, 5-year, 1-day rainfall excess and proportion of urban area | | Lag time, a function of basin ratio |
| Percentage runoff | A function of a soil index, a catchment wetness index, percentage of urban area and total rainfall depth | Separate allowance made for runoff from pervious and impervious areas; pervious area percentage runoff a function of a soil index, a catchment wetness index and total rainfall depth; 70% runoff allowed from impervious area (or 0·3 × proportion of urban area) | |
| Baseflow | ← Function of catchment wetness index and 5-year, 1-day rainfall excess → | | |

guide to the design of flood storage ponds compiled under the auspices of the Construction Industry Research and Information Association (Hall and Hockin, 1980), which incorporates the dimensionless unit hydrograph derived by Hall (1981). Both of these methods were based upon the design hydrograph method contained in the United Kingdom Flood Studies Report (Natural Environment Research Council, 1975), but differ in detail both from the latter and between themselves. The key features of all three methods are summarised in Table 8.1. Despite the differences in the synthetic unit hydrographs contained in FSSR5 and the CIRIA Guide, Packman (1981) has shown that both methods yield similar results.

## REFERENCES


Anderson, D. G. (1970) Effects of urban development on floods in northern Virginia. US Geol. Survey, Water-Supply Paper 2001-C, 22 pp.

Beard, L. R. and Chang, S. (1979) Urbanisation impact on streamflow. *Proc. Am. Soc. Civ. Engrs., J. Hydraul. Div.*, **105**(HY6), 647–59.

Bleek, J. M. (1975) Synthetic unit hydrograph procedures in urban hydrology. In *Proc. Nat. Symp. on Urban Hydrology and Sediment Control*, Kentucky Univ. Off. Res. Engng. Serv. Bull. No. 109, pp. 149–59.

Carter, R. W. (1961) Magnitude and frequency of floods in suburban areas. US Geol. Survey, Prof. Paper 424-B, pp. B9–B11.

Chang, M. and Lee, R. (1974) Objective double-mass analysis. *Wat. Resour. Res.*, **10,** 1123–6.

Crawford, N. H. and Linsley, R. K. (1966) Digital simulation in hydrology: Stanford Watershed model IV. Stanford Univ., Dept. Civ. Engng., Tech. Rept. No. 39, 210 pp.

Crow, E. L., Davis, F. A. and Maxfield, M. W. (1960) *Statistics manual* (Dover Publications, New York) 288 pp.

Cruise, J. F. and Contractor, D. N. (1980) Unit hydrographs for urbanising watersheds. *Proc. Am. Soc. Civ. Engrs., J. Hydraul. Div.*, **106**(HY3), 440–5.

Eagleson, P. S. (1962) Unit hydrograph characteristics for sewered areas. *Proc. Am. Soc. Civ. Engrs., J. Hydraul. Div.*, **88**(HY2), 1–25.

Edson, C. G. (1951) Parameters for relating unit hydrographs to watershed characteristics. *Trans. Am. Geophys. Un.*, **32,** 591–5.

Espey, W. H. and Winslow, D. E. (1974) Urban flood frequency characteristics. *Proc. Am. Soc. Civ. Engrs., J. Hydraul. Div.*, **100**(HY2), 279–93.

Espey, W. H., Morgan, C. W. and Masch, F. D. (1965) A study of some effects of urbanisation on storm run-off from a small watershed. Univ. of Texas, Austin, Center for Res. in Wat. Resour., Tech. Rept. HYD 07-6501.CRWR-2, 109 pp.

Espey, W. H., Winslow, D. E. and Morgan, C. W. (1969) Urban effects on the unit hydrograph. In Moore, W. L. and Morgan, C. W. (eds), *Effects of watershed changes on streamflow*. Proc. Wat. Resour. Symp. No. 2, Univ. of Texas, Austin, Centre for Res. in Wat. Resour. (Univ. of Texas Press, Austin), pp. 215–28.

FRANKE, O. L. (1968) Double-mass-curve analysis of the effects of sewering on groundwater levels on Long Island, New York, US Geol. Survey, Prof. Paper 600-B, pp. B205–B209.

GARBER, M. S. and SULAM, D. J. (1976) Factors affecting declining water levels in a sewered area of Nassau County, New York. *J. Res. US Geol. Survey*, **4,** 255–65.

GRAY, D. M. (1961) Synthetic unit hydrographs for small watersheds. *Proc. Am. Soc. Civ. Engrs., J. Hydraul. Div.*, **87**(HY4), 33–54.

GUNDLACH, D. L. (1978) Adjustment of peak discharge rates for urbanisation. *Proc. Am. Soc. Civ. Engrs., J. Irrig. Drain. Div.*, **104**(IR3), 325–9.

HALL, M. J. (1974) Synthetic unit hydrograph technique for the design of flood alleviation works in urban areas. In *Design of water resources projects with inadequate data*, Proc. Madrid Symp., UNESCO Studies and Reports in Hydrol., No. 16, Vol. 2, pp. 485–500.

HALL, M. J. (1977) The effect of urbanisation on storm runoff from two catchment areas in North London. Int. Assoc. Hydrol. Sci., Publ. No. 123, pp. 144–52.

HALL, M. J. (1981) A dimensionless unit hydrograph for urbanising catchment areas. *Proc. Instn. Civ. Engrs., Part 2*, **71,** 37–50.

HALL, M. J. and HOCKIN, D. L. (1980) Guide to the design of storage ponds for flood control in partly urbanised catchment areas. Construction Industry Research and Information Association, Tech. Note 100, 103 pp.

HARRIS, E. E. and RANTZ, S. E. (1964) Effect of urban growth on streamflow regimen of Permanente Creek, Santa Clara County (California). US Geol. Survey, Water-Supply Paper 1591-B, 18 pp.

HOLLIS, G. E. (1975) The effect of urbanisation on floods of different recurrence interval. *Wat. Resour. Res.*, **11,** 431–5.

INSTITUTE OF HYDROLOGY (1979) Design flood estimation in catchments subject to urbanisation. Institute of Hydrology, Wallingford, Flood Studies Supplementary Rept. No. 5, 18 pp.

JACKSON, T. J., RAGAN, R. M. and FITCH, W. N. (1977) Test of Landsat-based urban hydrologic modelling. *Proc. Am. Soc. Civ. Engrs., J. Wat. Resour. Plan. and Man. Div.*, **103**(WR1), 141–58.

JAMES, L. D. (1965) Using a digital computer to estimate the effects of urban development on flood peaks. *Wat. Resour. Res.*, **1,** 223–34.

JOHNSTONE, D. and CROSS, W. P. (1949) *Elements of applied hydrology* (Ronald Press, New York) 276 pp.

KOHLER, M. A. (1949) On the use of double-mass analysis for testing the consistency of meteorological records and for making required adjustments. *Bull. Am. Met. Soc.*, **5,** 188–9.

LINSLEY, R. K. (1943) Application of the synthetic unit-graph in the western mountain states. *Trans. Am. Geophys. Un.*, **24,** 580–7.

MARTENS, L. A. (1968) Flood inundation and effects of urbanisation in metropolitan Charlotte, North Carolina. US Geol. Survey, Water-Supply Paper 1591-C, 60 pp.

NASH, J. E. (1957) The form of the instantaneous unit hydrograph. Int. Assoc. Sci. Hydrol., Publ. No. 45, pp. 114–21.

NATURAL ENVIRONMENT RESEARCH COUNCIL (1975) *Flood Studies Report*, Vol. I: *Hydrological studies* (NERC, London) 550 pp.

NELSON, T. L. (1970) Synthetic unit hydrograph relationships, Trinity River tributaries, Fort Worth–Dallas urban area. In *Proc. Seminar on Urban Hydrol.*, Paper 6, US Army Corps of Engrs., Hydrologic Engng. Center, 18 pp.

PACKMAN, J. C. (1981) Effects of catchment urbanisation on flood flows. In *Flood Studies Report—Five Years on* (Thomas Telford Ltd, London) pp. 121–9.

RAO, A. R., DELLEUR, J. W. and SARMA, B. S. P. (1972) Conceptual hydrologic models for urbanising basins. *Proc. Am. Soc. Civ. Engrs., J. Hydraul. Div.*, **98**(HY7), 1205–20.

SAWYER, R. M. (1963) Effect of urbanisation on storm discharge and ground-water recharge in Nassau County, New York. US Geol. Survey, Prof. Paper 475-C, pp. C185–C187.

SEARCY, J. K. and HARDISON, C. H. (1960) Double-mass curves. US Geol. Survey, Water-Supply Paper 1541-B, 66 pp.

SINGH, R. (1968) Double-mass analysis on the computer. *Proc. Am. Soc. Civ. Engrs., J. Hydraul. Div.*, **94**(HY1), 139–42.

SNYDER, F. F. (1938) Synthetic unit hydrographs. *Trans. Am. Geophys. Un.*, **19,** 447–54.

VAN SICKLE, D. (1969) Experience with the evaluation of urban effects for drainage design. In Moore, W. L. and Morgan, C. W. (eds), *Effects of watershed changes on streamflow.* Proc. Wat. Resour. Symp. No. 2, Univ. of Texas, Austin, Center for Res. in Wat. Resour. (Univ. of Texas Press, Austin) pp. 229–54.

# 9

# Design of Stormwater Drainage Systems

## 9.1. INTRODUCTION

When a catchment area is urbanised and the amount of impervious cover in the form of roofs, roads and pavements increases, the need inevitably arises for the natural drainage network to be supplemented or even replaced completely by man-made systems of pipes and paved gutters. These systems of pipes or sewers generally assume a dendritic form in plan, similar to that of a network of natural channels. However, the hydrological design problems associated with sewerage, i.e. systems of sewers, differ from those concerned with channel works in that no measurements of surface water runoff are possible prior to construction. Design flood estimates for sewers must therefore be inferred from rainfall statistics using deterministic methods. As a corollary to this design approach, the performance of a sewerage system once constructed is rarely recorded unless problems are encountered with its behaviour under conditions approaching those of the design storm. This lack of incentive, coupled with the difficulties of gauging flows in sewers, has resulted in a dearth of flow records from sewered catchment areas, which has perhaps provided the biggest obstacle to the development of stormwater drainage design methods throughout their long history.

Sewerage systems may be broadly classified into two types, namely:

(1) combined systems, in which both the stormwater drainage and the domestic waste or sewage are conveyed in the same pipe network; and
(2) separate systems, in which the foul drainage is conveyed to the nearest treatment plant and the stormwater drainage is carried to the nearest watercourse in its own system of sewers.

In practice, partially combined systems are to be found, which carry the domestic sewage from only a proportion of the area. However, the stormwater discharge can be at least two orders of magnitude larger than the so-called dry-weather flow, and therefore provides the more dominant design consideration.

The flood estimation methods that have been applied to the design of stormwater drainage systems may be considered to fall into two broad categories: those which produce only an estimate of the peak flow rate, and the more comprehensive approaches that also provide the shape of the runoff hydrograph. These methods are described in Sections 9.2 and 9.3 respectively. With the wider availability of digital computers, the design hydrograph methods have increased in their scope and complexity. These later developments, which are distinguished primarily by the separate modelling of the above-ground and the below-ground phases of runoff, are discussed in Section 9.4.

The problems of urban drainage design can range from the analysis of existing sewer networks to the design of entirely new sewerage systems, and the area served may vary in size from a small housing estate to a large conurbation. In order to cover the range of possibilities, a design procedure incorporating a hierarchy of methods is required, similar to that for the estimation of floods on natural catchment areas presented schematically in Fig. 4.5. The application of this concept to stormwater drainage design is conveniently illustrated by the Wallingford Procedure, the details of which are outlined in Section 9.5.

## 9.2. MAXIMUM DISCHARGE METHODS

Until the early years of the twentieth century, urban drainage systems in the United Kingdom were designed on the basis of an average rainfall intensity which was assumed to be independent of duration. Even the main interceptor sewers running alongside the north and south banks of the River Thames in London were designed by Sir Joseph Bazalgette to criteria of this type. However, with the publication by the British Rainfall Organisation of statistical summaries of heavy falls of rain in short periods from 1888 onwards, the inverse relationship between the average rate of rainfall and duration became well established by observation. During the same period the first steps were taken to place urban drainage design on a more scientific footing, with measurements of rainfall and discharge from sewered catchment areas being undertaken by Kuichling (1889) in the United States and Lloyd-Davies (1906) in England. These two studies were similar in form, with flow rates being computed from records of depths at the outfall sewer and velocities estimated using a uniform flow formula. Despite the crudity of these experiments by modern standards of hydrometry, both authors were able to draw upon sufficient data to establish the basis of what is perhaps the most widely known of the simpler flood estimation techniques. Although Dooge (1957) has pointed out that the principles of the Rational Method were clearly expounded by Mulvaney (1851), the latter account was concerned with the arterial drainage of agricultural areas and credit for its introduction to sewerage design must remain with Kuichling and Lloyd-Davies.

The Rational Method presumes the existence of a time of concentration for every drainage area, which is defined as the time taken for flow from the most remote point in the catchment to reach its outfall. The peak discharge, $Q_p$, is then assumed to occur when the whole of the drainage area contributes to the flow, i.e. after an interval from the beginning of rainfall equal to the time of concentration. The magnitude of the peak flow rate is taken to be proportional to the effective rainfall or rainfall excess, i.e. the total rainfall minus the losses, during the time of concentration, giving rise to the equation

$$Q_p = 2{\cdot}78CiA \tag{9.1}$$

where the units of the discharge are litres/s, $A$ is the total catchment area above the outfall (ha), $i$ is the average rainfall rate (mm/h) during the time of concentration, $T_c$, and $C$ is a dimensionless runoff coefficient having a value less than unity.

When the Rational Method is applied to the design of urban drainage systems, the time of concentration is normally estimated from the sum of the time of flow in the sewer and a time of entry. The time of flow is usually computed on the assumption that the velocity of flow will be constant and equal to the value attained when the sewer is running full. The time of entry is an allowance for the time taken for water from the most remote point in the subcatchment of the pipe to flow into the nearest road gulley, and for urbanised areas has generally been taken to be between 1 and 3 min. As defined, the time of concentration is not therefore a measurable physical quantity. However, the assumption of a constant flow velocity permits the synthesis of a runoff hydrograph which serves to clarify the points upon which the Rational Method fails to describe observable hydrological phenomena.

For the purposes of this illustration, the catchment area is assumed to be rectangular in plan and sloping only in the direction of its outfall. Given the constant velocity assumption, the runoff hydrograph from this catchment will rise at a uniform rate from the beginning of the rainfall and reach the peak discharge, $q = Q_p/A$, at time $T_c$ (see Fig. 9.1(a)). The constant flow velocity assumption also implies that the time interval which elapses after the rainfall ceases before the flow falls to zero is also equal to $T_c$. The runoff hydrograph therefore assumes the trapezoidal shape shown in Fig. 9.1(a).

In practice, even with this simple geometry of catchment area, the runoff process is demonstrably more complex, as shown by the results from numerous laboratory catchment studies (for example, Izzard, 1946; Woo and Brater, 1962; Yu and McNown, 1964; Yoon and Wenzel, 1971; Shen and Li, 1973; Muzik, 1974). If the catchment surface is pervious, the initial soil moisture deficit must be satisfied

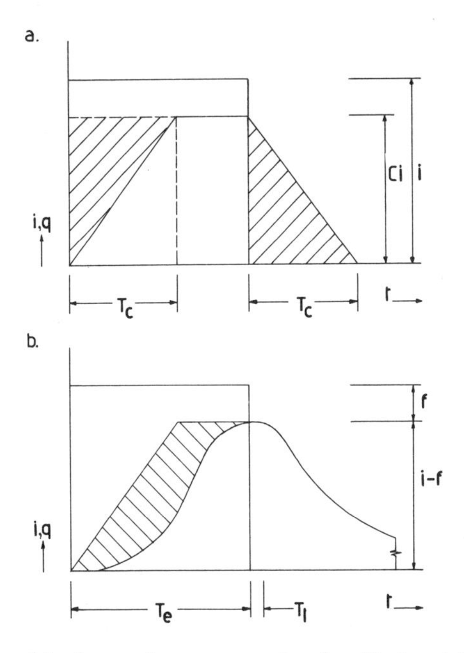


Fig. 9.1. (a) Flood hydrograph constructed using Rational Method assumptions. (b) Comparison between natural and Rational Method hydrographs.

along with depression storage before runoff can begin. The infiltration continues, but at a decreasing rate, $f$, throughout the storm. Surface retention causes further retardation of flow so that the runoff rate approaches $i-f$ asymptotically (see Fig. 9.1(b)). Assuming that the rainfall ceases when an equilibrium state is achieved at time $T_e$, there will be a short interval $T_1$ at or near the peak rate of flow caused by momentum effects before the recession begins. The falling limb is typically exponential in form as the runoff rate decreases, with a much longer duration than the rising limb.

A comparison between the Rational Method hydrograph of Fig. 9.1(a) and the natural hydrograph of Fig. 9.1(b) shows that a reasonable description of the time distribution of runoff is possible, but at the expense of underestimating the amount of water taken into storage during the rising limb. Since the volume of storage available in a sewerage system is generally much smaller than that of a natural channel network in a rural area of comparable size, the hatched area in Fig. 9.1(b) could be quite small. The estimates of the peak flow rate provided by the Rational Method might then be reasonable, provided that a representative value of the runoff coefficient was also chosen. According to eqn (9.1), the runoff coefficient is strictly defined as the ratio of the peak rate of runoff to the average rate of rainfall during the time of concentration. However, reference to Fig. 9.1(b) shows that, since the two hatched areas are equal and $q=Ci$, the runoff coefficient may also be defined as the ratio of the total volume of runoff to the total volume of rainfall (Williams, 1950). This duality has perhaps contributed towards some of the confusion over the choice of runoff coefficients for design purposes commented upon by Ardis *et al.* (1969), among others. The 'impermeability factors', of which extensive tabulations are to be found in many engineering handbooks (Chow, 1964; Water Pollution Control Federation and American Society of Civil Engineers, 1969), appear to be treated more as volumetric ratios. For catchment areas with a mixed land use, weighted average coefficients are computed.

Alternatively, overall impermeability factors have been expressed as a function of population density or building density (Meek, 1928; Roseveare, 1930; Escritt, 1965). Two typical relationships between the impermeability factor and the number of houses/ha are shown in Fig. 9.2 along with data obtained from a survey of a new town area in south-east England. The latter points show distinctly different trends according to the age of the development, thereby indicating the need

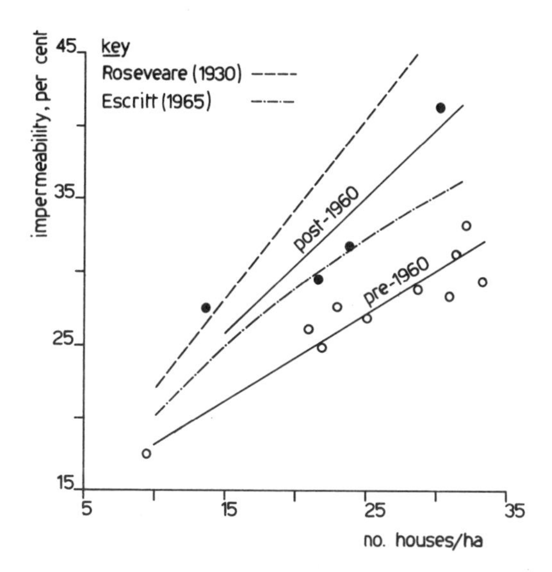


FIG. 9.2. Impermeability as a function of number of houses/ha for an English new town area (modified from Hall and Hockin, 1980, by permission of CIRIA).

for caution in applying the generalised relationships. Nevertheless, sample surveys or the analysis of aerial photographs can be costly and time-consuming, and for planning-level studies Gluck and McCuen (1975) have recommended the use of prediction equations in which the independent variables are parameters taken from census summaries. Those authors quote an example in which the variables employed were the population density and the distance from the centre of the down-town business district.

In effect, the impermeability factor is related to the amount of paved surface within an area having a particular land use. Lloyd-Davies (1906) appears to have been among the first to make the obvious simplifying assumption that the ratio of the paved to the total area can be taken as a representative runoff coefficient for an urban area. (Indeed, a distinction is sometimes drawn between the 'Lloyd-Davies Method', based upon this assumption, and the 'Rational Method', which employs weighted average impermeability factors.) Using data obtained by gauging four areas in Manchester, Meek (1928) was able to confirm the general validity of the Lloyd-Davies assumption. However, for more frequent rainfalls than the design event, the results obtained appeared to indicate that the catchment losses were constant. An explanation of this behaviour was subsequently offered by Appleby (1937), who drew attention to two aspects of the runoff coefficient: the ratio of paved to total area, $D$, and the capacity of the paved surfaces to absorb and retain water in terms of a constant soakage rate, $s$. The effective rainfall rate is therefore $i-s$, and the runoff coefficient may be expressed as

$$C=D(i-s)/i=D(1-[s/i]) \tag{9.2}$$

a form of relationship which was able to explain Meek's results. The average soakage rate was found by Appleby to be 1·8 mm/h.

The adequacy of the proportion of paved area as an approximation to the runoff coefficient was also examined by Watkins (1962), using data from an extensive gauging programme on urban catchment areas in England undertaken between 1953 and 1958. For 104 storms, the average ratio of total volume of runoff to total volume of rainfall was 0·79, and only 7 storms yielded ratios exceeding unity, when rainfall volume was calculated on the paved area only. The approximation was therefore considered adequate for design purposes.

Since the product $CA$ in eqn (9.1) is assumed to be constant for a given catchment area, the statistical properties of the average rainfall rate are transferred directly to the peak discharge, i.e. the $T$-year rainfall intensity yields the $T$-year flood. This equality was used by Schaake *et al.* (1967) to test the validity of the Rational Method on 20 urban areas that had been gauged as part of the Johns Hopkins University Storm Drainage Research Project (see Schaake, 1969). In the majority of cases, Schaake *et al.* found that, on comparing the frequency distributions of peak runoff rates and average rainfall intensities for each catchment, the equality of return periods assumption was reasonably valid, thereby lending credence to the Rational Method as a simple design tool.

The simplicity of design techniques such as the Rational Method can be deceptive. In a survey of storm drainage design practices within the State of Wisconsin, Ardis *et al.* (1969) found that only 6 out of 23 design offices were applying the method correctly. The principal sources of error were the computation of runoff coefficients, which has already been discussed above, and the selection of the appropriate average rainfall rate.

Since the design of an urban drainage system proceeds downstream from the watershed, the time of concentration increases as further sub-areas are included in the calculations. For any given return period, the average rainfall intensity decreases as the duration of rainfall increases (see Chapter 3). The design rainfall rate therefore decreases as the drainage network is extended downstream. With the times of concentration of 5–30 min that are typical of many urban sewerage systems, the variations with duration of average rainfall intensities are sufficient to lead to a markedly different series of pipe sizes than the constant intensity assumed by 16 of the respondents to the survey by Ardis *et al.* (1969).

The Rational Method is also known to yield erroneous results under certain design conditions. In particular, for drainage systems in which the contributing area does not increase uniformly with time, the highest peak runoff rate may be produced by a design storm whose duration is *less* than the time of concentration. The occurrence of this anomaly may be illustrated by an analysis of the drainage area sketched in Fig. 9.3. Assuming that the area between $O$ and $X$ is $A_1$ and its time of concentration is $T_1$, and that between $O$ and $Y$ the area is $A_2$ with a time of concentration of $T_2$, application of eqn (9.1) to both areas yields the following ratio of the design discharges from areas 1 and 2:

$$Q_1/Q_2 = i_1 A_1 / i_2 A_2 = (A_1/A_2)/(i_2/i_1) \qquad (9.3)$$

where the runoff coefficient is assumed to be uniform. Since $T_2$ is greater than $T_1$, $i_2$ must be less than $i_1$. Therefore, the ratio $A_1/A_2$ may exceed the ratio $i_2/i_1$, whereupon $Q_1$ will exceed $Q_2$.

In summary, the Rational Method does not take into account

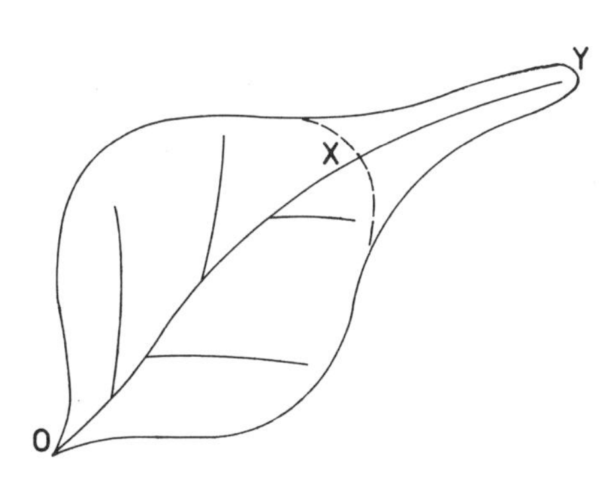


FIG. 9.3. Catchment illustrating the importance of rate of increase in contributing area to the Rational Method.

variations in time of:

(i) rainfall intensity;
(ii) flow velocity;
(iii) temporary storage in the sewer system; and
(iv) the rate of increase in the contributing area.

According to Watkins (1962), the Rational Method only provides reasonable estimates of peak discharge when the errors resulting from items (iii) and (iv) are compensating. The same author concluded that the Rational Method is only suitable for design purposes when the drainage areas are sufficiently small for pipe diameters not to exceed 610 mm. However, the latter stipulation is conditioned more by the increments in the available commercial pipe sizes within this range of diameters, where one change in size of 75 mm may double the carrying capacity. Safety is therefore bought relatively cheaply.

Attempts to refine the flood estimates provided by the Rational Method have concentrated largely on item (iv) above, with the use of a plot showing the variation with time from the beginning of the storm of the area of the catchment contributing to the flow at the outfall, referred to as a time–area diagram. In order to prepare the time–area diagram, the catchment is divided into sub-areas that are small enough for contributing area to be directly proportional to time during the time of concentration. Given these triangular-shaped diagrams for each sub-area, the time–area diagram of the whole catchment is constructed by linear superposition, with the starting points of the sub-area diagrams being staggered according to times of flow to the catchment outfall. Figure 9.4 illustrates the construction of a composite time–area diagram for two areas, $A_1$ and $A_2$, having times of concentration $T_1$ and $T_2$, with their outfalls a time of flow, $T_3$, apart. The diagram is referred to the downstream outfall of area $A_1$.

Figure 9.4 shows that until time $T_3$, only area $A_1$ contributes to flow. By time $T_1$, area 1 is contributing fully, but area 2 has only just begun, making the total area

$$A_1 + A_2(T_1 - T_3)/T_2$$

At time $T_2 + T_3$, which is the time of concentration of the composite area $A_1 + A_2$, the whole catchment is contributing in accordance with Rational Method assumptions.

The time–area diagram has formed the basis for two distinct types of methods, the first of which may be referred to as the Tangent Methods.

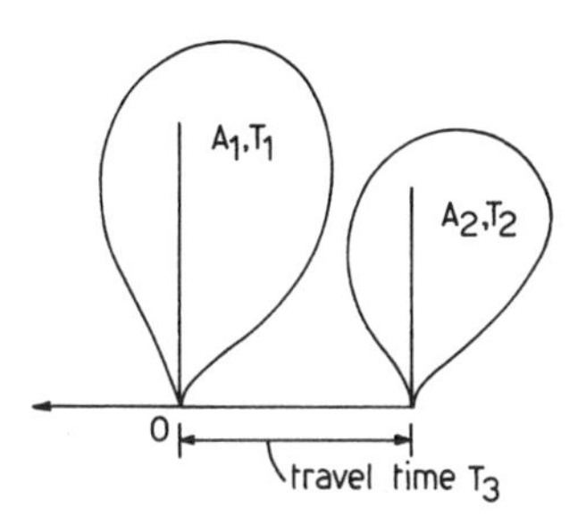


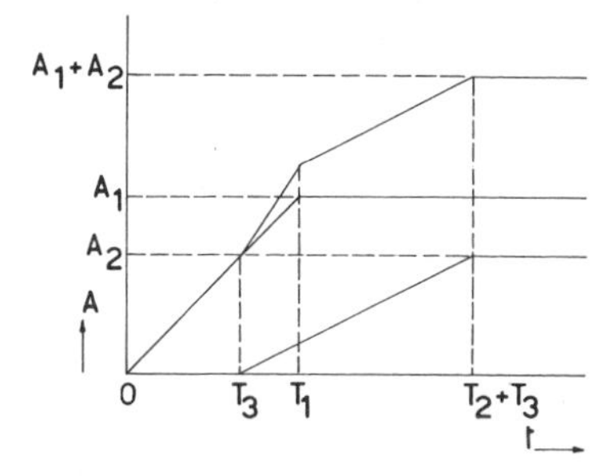


FIG. 9.4. Construction of a time–area diagram.

As their name implies, this group involves a geometrical construction, the basis of which is shown in Fig. 9.5. The construction begins from a point $P$ located at a distance $b$ to the left of the origin of the time–area diagram. If an arbitrary line is drawn from $P$ to cut the time–area diagram at point $N$, whose coordinates are area $A'$ and time $T'$, eqn (9.1) may be applied to compute

$$Q_N = CiA' \tag{9.4}$$

Substituting a simple rainfall intensity–duration relationship into eqn (9.4) using the duration $T'$, $i = a/(T'+b)$, where $a$ and $b$ are constants which vary with return period (see Chapter 3),

$$Q_N = CA'a/(T'+b) \tag{9.5}$$

However, Fig. 9.5 shows that $A'$ is the vertical of a right-angled triangle whose base is of length $T'+b$, so that eqn (9.5) reduces to

$$Q_N = Ca \tan \phi \tag{9.6}$$

Since both $C$ and $a$ are constants, the discharge is a maximum when

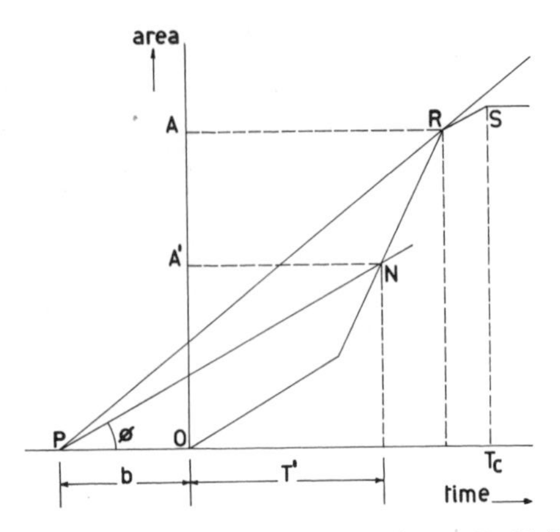


FIG. 9.5. The Tangent Method according to Reid (1927).

$\tan \phi$ is a maximum, i.e. when the line $PN$ is tangential to the time–area diagram, as shown by the line $PR$.

Although the time–area diagram provides information on the configuration of the drainage area, the construction of the tangent as described above merely locates the point $R$ without making use of this additional knowledge. If the points $R$ and $S$ coincide, the Tangent Method will yield an identical result to the Rational Method. However, if the point of maximum convexity is nearer to the origin, the reduction in the abscissa is larger than that in the ordinate. Since rainfall intensity increases rapidly with decreasing duration, the maximum discharge computed from the coordinates of the point $R$ can only exceed the Rational Method estimate (see Watkins, 1962).

As presented in this form by Reid (1927), the Tangent Method does not allow for the possibility that the peak discharge could occur after the area around the outfall has ceased to contribute. In order to take the latter circumstances into account, Riley (1932) suggested moving the origin $O$ to the point of maximum concavity in the time–area diagram. The point $P$ is then located a horizontal distance $b$ to the left as before. A more formal version of this approach, in which a second time–area diagram was replotted a distance $b$ to the left of the first and a tangent drawn to touch both curves, was presented by Norris (1946). The coordinates of the tangent point $R$ are then measured from the point of maximum concavity, as shown in Fig. 9.6. The deficiencies of the Reid method are equally evident in this modified approach. If the

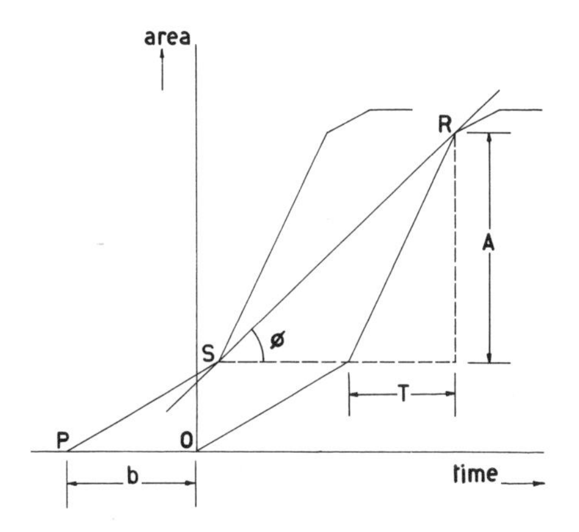


FIG. 9.6. The Modified Tangent Method according to Norris (1946).

points $P$ and $S$ in Fig. 9.6 coincide, the modified and the single Tangent Methods will furnish identical flood estimates. Otherwise, the double Tangent Method will yield the larger estimates in all cases.

Another Modified Tangent Method has been suggested by Escritt (1965) which involves the preparation of a transparent overlay containing curves of equal runoff. The latter are obtained by assuming a discharge, $Q_p$, and then dividing this value by a series of rainfall intensities, $i$, so that the impermeable areas, $CA$, corresponding to a series of durations can be computed. Since Escritt employed a rainfall intensity–duration relationship of the form $i = a/T^m$ instead of $i = a/(T+b)$, the transparent overlay simply provides a family of curvilinear tangents, which are positioned over the time–area diagram and adjusted until the maximum discharge is obtained. This elaborate procedure does not alter any of the basic assumptions behind the Tangent Method, and the bias in the direction of larger discharges of all its variants has resulted in their abandonment in favour of more flexible alternatives.

The second group of methods which employs the time–area diagram, known as the Typical Storm Methods, differs from the Tangent Methods in producing a runoff hydrograph and not just an estimate of the peak flow rate. Before discussing the Typical Storm Methods, however, the following connection between the Rational Method and the Unit Hydrograph Method may be noted. Following Nash (1958), eqn (9.1) may be rewritten in the form

$$Q = CA(r/T_c) = (CA/T_c)\int_{t-T_c}^{t} i(\tau)\,d\tau \qquad (9.7)$$

where $r$ is the total rainfall depth and is expressed as the time integral of the instantaneous rainfall intensity, $i(\tau)$. Equation (9.7) may be compared with the convolution integral

$$Q = \int_0^t u(0, t-\tau)i(\tau)\,d\tau \qquad (9.8)$$

where $u(0, t-\tau)$ is the instantaneous unit hydrograph, to demonstrate that the Rational Method is a special case of the Unit Hydrograph Method in which

$$u(0, t-\tau) = (CA/T_c); \qquad 0 \leqslant \tau \leqslant T_c \qquad (9.9)$$

is a rectangular pulse of constant height and duration $T_c$.

## 9.3. DESIGN HYDROGRAPH METHODS

The development of flood hydrograph estimation methods for sewerage systems may be considered to consist of two separate phases, the first of which began with the so-called Typical Storm Methods. The latter differ primarily in the distribution of rainfall which is assumed for a specified return period. As described by Ross (1921), the method involves the drawing of isochrones, i.e. lines of equal travel time, on a map of the drainage area using a time increment, $\Delta t$, such that $T_c = m\,\Delta t$ with $m$ an integer. The areas between adjacent isochrones, $a_j$, $j = 1, 2, \ldots, m$, are then measured. Assuming that the storm profile consists of a series of average rainfall intensities, $i_1$, $i_2$, $i_3$, . . . , within successive time increments of $\Delta t$, the ordinates of the discharge hydrograph may be written as

$$\begin{aligned} Q_1 &= Ci_1A_1 \\ Q_2 &= Ci_1A_2 + Ci_2A_1 \\ Q_3 &= Ci_1A_3 + Ci_2A_2 + Ci_3A_1 \qquad (9.10) \\ &\ldots \end{aligned}$$

where $C$ is the runoff coefficient of the drainage area.

Ross (1921) suggested that the storm profile should be constructed with the average rainfall rates within the isochrone intervals, $\Delta t$,

$2\,\Delta t, \ldots, m\,\Delta t$, being read from the rainfall intensity–duration relationship for the same return period. As noted in Chapter 3, the storm profile which results from this approach has its maximum ordinate within the first time increment and a series of diminishing rainfall intensities in the subsequent intervals. The return period attributed to this profile was the same as that of the rainfall intensity–duration relationship from which it was derived.

Subsequent authors, notably Hawken (1921) and Judson (1933), criticised the sequence of decreasing rainfall intensities, not because of the arbitrary shape, but because such a series may not yield the maximum discharge when applied in eqns (9.10). A rearrangement of the rainfall intensities was suggested such that the largest would be paired with the largest area increment, the second largest with the second largest area increment, and so on. The even more implausible shapes of storm profile which resulted from this exercise were also allocated the same return period as the basic rainfall intensity–duration relationship.

With the increasing use of autographic raingauges in the United Kingdom from 1930 onwards, storm profiles could be based upon observed events. However, with the exception of Coleman and Johnson (1932), who were unable to define the frequency of the chosen design storm in anything but the broadest of terms, many writers continued to advocate the use of synthetic storm profiles. One of the more widely known methods, introduced by Ormsby (1933) and modified by Hart (1933), employed two storm profiles, one peaking at mid-duration and the other at one-third of the duration. Both were envelope curves, in the sense that the average rainfall intensities for all time intervals less than or equal to the storm duration corresponded to the same return period (see Chapter 3). Although a distinct improvement on the arbitrary rearrangements of average rainfall intensities proposed by Ross (1921), Hawken (1921), Judson (1933) and others, the Ormsby–Hart profiles again undoubtedly represented a rarer event than the rainfall intensity–duration–frequency relationship upon which they were based.

The Typical Storm Methods all assume that the discharge at the outfall is composed of increments in discharge, $\mathrm{d}Q$, obtained from an average rainfall intensity, $i(\tau)$, falling on an increment in impervious area, $\mathrm{d}A$, with a time of flow to the outfall of $t-\tau$:

$$\mathrm{d}Q_t = i(\tau)\,\mathrm{d}A_{t-\tau} = i(\tau)(\mathrm{d}A/\mathrm{d}\tau)_{t-\tau}\,\mathrm{d}\tau \qquad (9.11)$$

The total discharge, $Q$, may therefore be written as

$$Q = \int_0^t (\mathrm{d}A/\mathrm{d}\tau)_{t-\tau} i(\tau)\, \mathrm{d}\tau \tag{9.12}$$

Comparing eqns (9.8) and (9.12) then shows that the Typical Storm Methods are a special case of the Unit Hydrograph Method in which

$$u(0, t-\tau) = (\mathrm{d}A/\mathrm{d}\tau)_{t-\tau} \tag{9.13}$$

i.e. the instantaneous unit hydrograph is given by the gradient of the time–area diagram, also known as the time–area concentration curve (Nash, 1958).

The second phase in the development of design hydrograph methods in the United Kingdom culminated in what is now known as the Transport and Road Research Laboratory (TRRL) Hydrograph Method. This approach was evolved during an extensive research programme conducted at the then Road Research Laboratory into design methods for the surface water drainage systems for large housing estates. Initially, only a rural catchment lying within the development area of a new town was monitored (see Watkins, 1956), but between 1953 and 1958 attention was turned to the gauging of 11 urban areas, which included a short length of road, 2 small factory areas, 4 small housing estates, 3 older built-up areas and a medium-sized housing development (Watkins, 1962, 1963).

The TRRL Hydrograph Method is a two-stage procedure, the first half of which is identical to that of the Typical Storm Methods and represented by eqns (9.10), with the runoff being assumed to occur only from the paved area within the catchment. The results obtained by applying this approach to 286 recorded storms showed that the computed discharge hydrographs consistently reached their peak too early and overestimated the observed maximum flow. This lack of agreement was attributed by Watkins (1962, 1963) to the constant velocity assumption implicit in the construction of the time–area diagram. In practice, depths of flow in a system of sewers change with variations in the rate of runoff, thereby altering the volume of water retained within the pipes. In effect, the sewer system behaves like a reservoir, and its action was therefore simulated by routing the unmodified (Typical Storm) hydrograph through a hypothetical storage having the same relationship between discharge and volume of water retained as the pipe network down to the point of interest.

This approach presupposes that the storage–discharge relationship or retention curve is unique for any given sewerage system. For existing pipe networks, the retention curve may be constructed from the recession curves of observed hydrographs, since the area under the curve to the right of each ordinate provides an estimate of the volume retained in the sewers at that flow rate. However, for systems under design, the retention curve had to be calculable from the dimensions of the pipe network, and the assumption was made that in a correctly designed system the proportional depth remained constant throughout the network at any given flow rate. This assumption has also to be applied when analysing existing systems for which no flow records are available, but experience has shown that, for pipe networks which do not decrease in diameter from the outfall to their upstream extremities, retention can be grossly overestimated. Particular care is therefore necessary in dealing with older types of development which can have exceptionally large upstream sewers.

The routing calculation is illustrated schematically in Fig. 9.7. Given an unmodified hydrograph with ordinates, $P_1, P_2, \ldots$, at intervals of $\Delta t$, the ordinates of the routed hydrograph, $Q_1, Q_2, \ldots$, at the same time intervals are obtained by approximating both hydrographs by a series of straight-line segments over successive $\Delta t$. Assuming that the storage during the first time increment is $R_1$, the geometry of these approximations gives

$$P_1(\Delta t/2) = Q_1(\Delta t/2) + R_1 \tag{9.14}$$

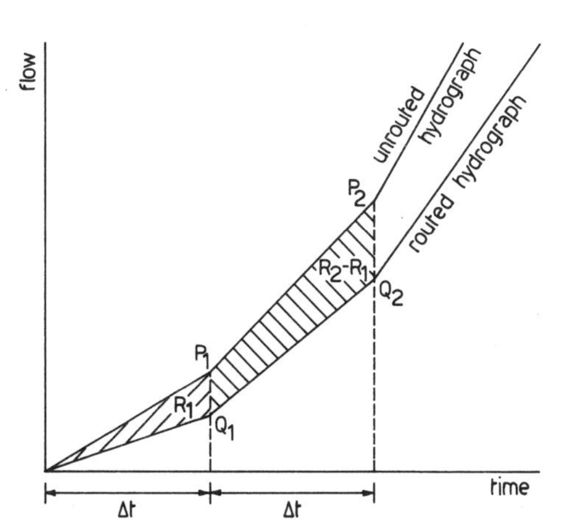


FIG. 9.7. Routing of the unmodified hydrograph in the TRRL Hydrograph Method (redrawn from Watkins, 1962, by permission of the Director, TRRL).

If the storage during the second time increment is given by $R_2 - R_1$,

$$(P_1 + P_2)(\Delta t/2) = (Q_1 + Q_2)(\Delta t/2) + (R_2 - R_1) \qquad (9.15)$$

which may be rearranged in the more convenient form

$$(P_1 + P_2 - Q_1)(\Delta t/2) + R_1 = Q_2(\Delta t/2) + R_2 \qquad (9.16)$$

such that both the unknowns at each time step are on the right-hand side of the equation. Since the $Q$–$R$ relationship is known, a graph of $Q$ versus $Q(\Delta t/2) + R$ may be prepared for ease in solving eqns (9.14) and (9.16).

Since the early 1960s, guidance on the selection of design methods for stormwater drainage systems in the United Kingdom has been available in Road Note 35 (Road Research Laboratory, 1963; Transport and Road Research Laboratory, 1976). This publication endorsed the conclusion drawn by Watkins (1962, 1963) that the Rational Method should only be employed on relatively small areas in which the sewers did not exceed 610 mm in diameter, and recommended the use of the TRRL Hydrograph Method both for larger catchments and for cases in which outflow hydrographs were required. Subsequent experience with the latter method showed that the redesign of existing sewerage systems was a more frequent requirement than the design of entirely new systems. The TRRL Hydrograph was therefore modified so that the constant proportional depth assumption was applied to individual pipe lengths, which were considered in turn beginning with the pipe furthest upstream (see Watkins and Young, 1965; Young, 1973). This change was not incorporated into Road Note 35 until the publication of the 2nd edition in 1976.

A survey carried out in 1975 (Hydraulic Design of Storm Sewers Working Party, 1976) showed that by then the methods recommended in Road Note 35 accounted for 96% of all storm drainage design work in the United Kingdom. However, the TRRL Hydrograph method has not been without its critics. For example, Escritt and Young (1963) contended that the time–area diagram implicitly allowed for storage in the sewerage system and that the routing of the unmodified hydrograph was tantamount to allowing for the same volume of storage twice. According to those authors, the attenuation observed in recorded storm hydrographs was attributable to surface storage.

These arguments serve to illustrate the problems of reconciling the assumptions implicit in simple design methods with knowledge of the rainfall–runoff process that they purport to simulate. In discussing such methods, tenuous physical analogies are perhaps better ignored and the basis of the approach regarded as a conceptual model of catchment behaviour. The form of the conceptual model underlying the TRRL Hydrograph Method has been amply justified by the comparison between the observed and reconstituted hydrograph carried out during the original study (Watkins, 1962, 1963), and by the encouraging results of independent trials of the method in the United States (Terstriep and Stall, 1969), Australia (Aitken, 1973, Heeps and Mein, 1973, 1974) and Canada (Marsalek *et al.*, 1975; James F. MacLaren Ltd, 1975). In addition, an appraisal by Colyer (1977) of the published information has shown that the TRRL Hydrograph Method appeared to be more reliable in simulating recorded flood hydrographs than several other computer-based procedures developed subsequently in other countries. However, these extensive trials of the method have shown that, other than in temperate climates, neglecting the runoff from the pervious areas within the catchment can lead to an underestimation of peak discharges. Watkins (1976) has since suggested an additional modification to the TRRL Hydrograph Method for tropical climates in which the runoff from the pervious portion of the catchment is modelled by means of a linear reservoir (see also Watkins and Fiddes, 1978).

The amount of calculation involved in the TRRL Hydrograph Method is large even for the simplest of sewerage systems and the use of a digital computer is necessary apart from the most trivial of schemes. Many local authorities in the United Kingdom have developed modifications to the original program to suit their particular requirements (see, for example, Cook and Lockwood, 1977). Versions of the program capable of dealing with pipes under surcharge have also been written (Martin and King, 1978; Thompson and Lupton, 1978). However, apart from the provision of such additional options, the availability of the computer appears to have had little influence on the structure of the method, the evolution of which may be traced over a period of more than 40 years. The subsequent phase in the development of stormwater drainage design methods has displayed an opposing tendency, with attention turning to more elaborate and physically plausible representations of the rainfall–runoff process. A particular feature of the new generation of design methods has been the separate consideration of the above-ground and below-ground components of the transformation of rainfall into discharge. The former is concerned with the synthesis of the surface runoff hydrograph at each stormwater inlet or road gulley, and the latter involves the routing of the gulley hydrographs through the sewer network.

## 9.4. DESIGN BY URBAN HYDROLOGICAL MODELLING

Among the first of the sewer design methods in which the above-ground and below-ground phases of the rainfall–runoff process were treated separately was the Chicago Hydrograph Method, a description of which was presented by Tholin and Keifer (1960). As its title implies, this method was devised specifically for use in the Chicago area where main sewers were generally laid out in parallel at intervals of about 0·8 km, and the lateral sewers served sub-areas of approximately 25 ha. The method was based upon a 3 h, once-in-5-year design storm derived from local rainfall records. The transformation of this storm into a runoff hydrograph began with the abstraction of infiltration losses to give a hydrograph of overland flow and depression storage supply. After deducting an allowance for depression storage, this hydrograph was routed through gutter storage to give the variation of flow with time at the nearest road gulley. The below-ground component of the rainfall–runoff model then began with the routing of the gulley hydrographs through the lateral sewer serving each sub-area. If the discharge hydrograph from a group of sub-areas was required, flows from the lateral sewers could then be routed down a main sewer.

The layout of both the lateral and main sewers and the uniformity of the sub-area geometry and land use served to reduce the amount of calculation involved in the Chicago Hydrograph Method. Even so, the approach was too laborious for hand calculations, and a digital computer was employed to produce a series of design charts which Tholin and Keifer claimed had made the method as straightforward in application as the Rational Method. However, with the wider availability of computer facilities, this intermediate step of preparing design charts has been superseded by more direct interactive computer use.

Perhaps the most widely known of the computer-based urban rainfall–runoff models, similar in form to that embodied in the Chicago Hydrograph Method, is the Storm Water Management Model (SWMM). This model is available in the form of a computer program containing over 10 000 FORTRAN statements, which was originally developed by a consortium of two firms of consulting engineers and a university under contract to the then US Federal Water Quality Administration (now the US Environmental Protection Agency). SWMM differs from the Chicago Hydrograph Method and its contemporaries in treating both the quality and quantity of urban runoff. Torno (1975) has described the computer program as consisting of four major blocks of subroutines controlled by a fifth group of executive routines. The RUNOFF block is concerned with the derivation of runoff hydrographs and their associated pollutant loadings. The TRANSPORT block routes both the hydrographs and the time variations of individual pollutants (also referred to as 'pollutographs' or 'chemographs') through the sewerage system. The STORAGE and RECEIV blocks simulate the action of a sewage treatment plant and the impact of discharges on the watercourse receiving the effluent respectively. The water quality models incorporated into SWMM, which have been discussed by Lager *et al.* (1971), are considered more fully in Chapter 11. The hydrological model contained within the RUNOFF block was described by Chen and Shubinski (1971).

The application of SWMM involves the division of the drainage area into a network of idealised elements, each of which consists of a rectangular plane with uniform land use, slope and surface characteristics. These planes need not be equal in size, but irregular shapes of sub-area must be approximated by rectangles of equivalent mean width. As described by Chen and Shubinski (1971), the overland flow hydrograph from each rectangular plane is derived from water balance computations at each step in which allowances are made for both infiltration and depression storage. These overland flow hydrographs are then routed through gutter storage. The subsequent routing of flows through the lateral sewers may be carried out either in the RUNOFF block using a simplified approach or, where backwater effects are likely to be significant, in the TRANSPORT block using a more sophisticated technique.

The University of Cincinnati Urban Runoff Model (UCURM) described by Papadakis and Preul (1972) is another computer-based model similar in form to SWMM. However, UCURM depends upon the division of the catchment into sub-areas that are either completely pervious or completely impervious. On each of the latter, allowances are made for infiltration, surface retention and overland and gutter flow before routing the gulley hydrographs through the pipe system. In its original form, UCURM was criticised by Heeps and Mein (1973, 1974) for its over-simplified approach to the treatment of infiltration and depression storage on pervious sub-areas, and the inefficiency of a solution procedure for the routing of overland flow. Replacement of the latter by an iterative technique achieved a remarkable reduction in computer central processor time to 4% of the original figure.

Although UCURM cannot be considered to have progressed very far beyond the initial development stage at the time of the study by Heeps and Mein (1973, 1974), the results obtained by the latter authors serve to illustrate the care which must be exercised in both formulating rainfall–runoff models and programming their solution. In this context, more attention appears to have been paid to the below-ground phase of runoff, where the relatively simple geometry of the pipe network makes the use of physically based models a practical possibility. Comparative studies by Yevjevich and Barnes (1970), Cunge (1974), Yen and Sevuk (1975) and Zaghloul (1977) among others have shown that simplified methods of routing are adequate for small pipe networks, but for larger systems, where backwater effects may be important, solution procedures based upon the full unsteady flow equations are necessary.

The above-ground phase of runoff is more complex, and in the absence of a simple sub-area geometry, as in the case of the Chicago Hydrograph Method, modellers have resorted to the use of idealised elements, such as rectangular planes of equivalent area, as in SWMM, or completely pervious and impervious subcatchments, as in UCURM. Some authors have adopted a systems approach in which, instead of attempting to model the above-ground phase of runoff in detail, the simulation of gulley hydrographs is accomplished by means of a simple conceptual model. For example, a single linear reservoir has been employed by Viessman (1966, 1968) and Watt and Kidd (1975). A single linear reservoir having different parameter values during wet and dry periods has been suggested by Swinnerton *et al.* (1973) for the estimation of runoff from motorways running in cuttings. More recently, Kidd and Helliwell (1977) have used a non-linear reservoir, i.e. a device in which storage is proportional to a power less than unity of its discharge, as the basis of a rainfall–runoff model for small urban areas. This model, further details of which have been supplied by Kidd (1978), has been incorporated into the hydrograph method forming part of the Wallingford Procedure described below in Section 9.5. This conceptual approach has been extended to the modelling of both above-ground and below-ground phases. The use of a single linear reservoir by Willeke (1966) demonstrated the inflexibility of a single-parameter model for this purpose. The cascade of two unequal linear reservoirs, i.e. two linear reservoirs with different storage constants in series, outlined by Sarginson and Bourne (1969), and the use of a linear reservoir and linear channel by Viessman *et al.* (1970) have both displayed greater potential.

The above discussion serves to illustrate the wide range of methods, many of which have been developed during the last decade, that is presently available to the designers of urban drainage systems. The choice of which method is the most appropriate among the several contenders is hardly straightforward.

Firstly, a clear distinction must be drawn between design methods and simulation methods. The former are able to calculate the pipe sizes required for a new sewerage system, given the design storm, the layout of the network and other descriptors of the catchment, whereas the latter analyse the performance of an existing system or an initial design produced by applying another method. Using these criteria, the design methods are limited to the Rational, TRRL and Chicago Hydrograph Methods and a development of the TRRL Method by Terstriep and Stall (1969, 1974) known as the Illinois Urban Drainage Area Simulator (ILLUDAS). The remainder are essentially simulation methods, which nevertheless are as useful in certain applications, such as the renovation and renewal of existing sewerage systems, as the design methods.

The second factor which may prevent a designer from making a clear choice between different design methods is a lack of information on their relative performance. A review by Colyer (1977) of the limited amount of comparative testing that has been conducted, including the studies by Papadakis and Preul (1973), Heeps and Mein (1973, 1974), Marsalek *et al.* (1975) and James F. MacLaren Ltd (1975) among others, has shown that each exercise has tended to be confined to a limited number of the available methods. Moreover, the results obtained showed that several of the methods were capable of simulating observed events to an accuracy approaching that of the recorded data. The emphasis on simulation in such testing is perhaps inevitable, since design methods can only be examined properly within a probabilistic framework, as exemplified by Schaake *et al.* (1967) in their evaluation of the Rational Method.

Given that several approaches are capable of providing a similar performance in simulation, the choice of a design method tends to depend upon other factors, such as the size and nature of the drainage area and the accuracy required. As Colyer and Pethick (1976) concluded from their comprehensive literature review, there is therefore no single 'best' method. For the smaller catchments, several approaches of comparable simplicity to the Rational Method have been developed but have attracted little support. With the 75 mm increment in pipe diameter currently available, the Rational Method continues to provide reasonable reliability and to achieve significant economies in design costs. For larger drainage areas, the potential savings in materials and construction costs are sufficient to justify a more sophisticated approach. The apparent equality in performance of several suitable methods inevitably forces the choice on to the availability of supplementary features, such as the treatment of surcharging or the modelling of water quality as well as quantity. Additional factors, such as the familiarity of the designer with the method, the availability of computer programs and support in the form of up-to-date documentation, all play their part in influencing a final decision.

In summary, an urban stormwater drainage design procedure must consist of several methods, each of which is appropriate to a particular range of drainage area and an acknowledged accuracy of peak flow estimation. The more complex the design problems, the more sophisticated the technique required to obtain their solution. Further subdivision of the procedure is possible according to the need for design or simulation, or the representation of ancillary structures such as stormwater overflows or detention tanks. This hierarchical approach to the design of surface water drainage systems is readily illustrated by the Wallingford Procedure (National Water Council and Department of the Environment, 1981).

## 9.5. THE WALLINGFORD PROCEDURE

The Wallingford Procedure for the design and analysis of urban storm drainage networks was based upon the results of a collaborative research programme carried out in the United Kingdom between 1974 and 1981 by the Hydraulics Research Station, the Institute of Hydrology and the Meteorological Office, and coordinated by the National Water Council/Department of the Environment Working Party on the Hydraulic Design of Storm Sewers. The Procedure consists of four methods:

1. The Wallingford Rational Method: a modified version of the

Rational Method as described in Section 9.2 intended for use on outline designs or on homogeneous areas of up to 150 ha; both manual and computer-based versions of this method are available, the latter including the facility to simulate stormwater overflows.

2. The Wallingford Hydrograph Method: a computer-based approach which models the above-ground and below-ground phases or runoff separately; this method may be employed for both design and simulation (see Price and Kidd, 1978), and allowances may also be made for the action of stormwater overflows, on-line and off-line detention tanks and pumping stations.
3. The Wallingford Optimising Method: a computer-based technique for obtaining the pipe diameter, depth and gradient associated with the minimum construction cost using the discrete differential dynamic programming technique (see Mays and Yen, 1975; Price, 1978).
4. The Wallingford Simulation Program: a computer-based method with which the performance of both an existing system and a proposed design may be examined under surcharged conditions (see Bettess *et al.*, 1978); stormwater overflows, on-line and off-line detention tanks and pumping stations may also be taken into account.

These methods may be applied to both separate and combined sewerage systems, although the calculation of foul sewage flows is not included. No allowances are made for the calculation of runoff from any rural areas that may contribute to an urban drainage network, and no water quality modelling is attempted. Nevertheless, the Procedure allows the hydraulic and cost consequences of alternative design standards relating to both the pipe network and any ancillary structures to be evaluated for a minimum expenditure of time and effort on the part of the designer.

The selection of the method most appropriate for a particular design requirement is assisted by following the flowchart presented in Fig. 9.8. For the design of new systems, the Modified Rational, Hydrograph or Optimising Methods can be employed. The flow calculations for the Optimising Method are carried out using the Modified Rational Method, and so the discharge estimates obtained from these approaches should be similar, unless the gradient optimisation substantially alters times of concentration. The Optimising Method may not be

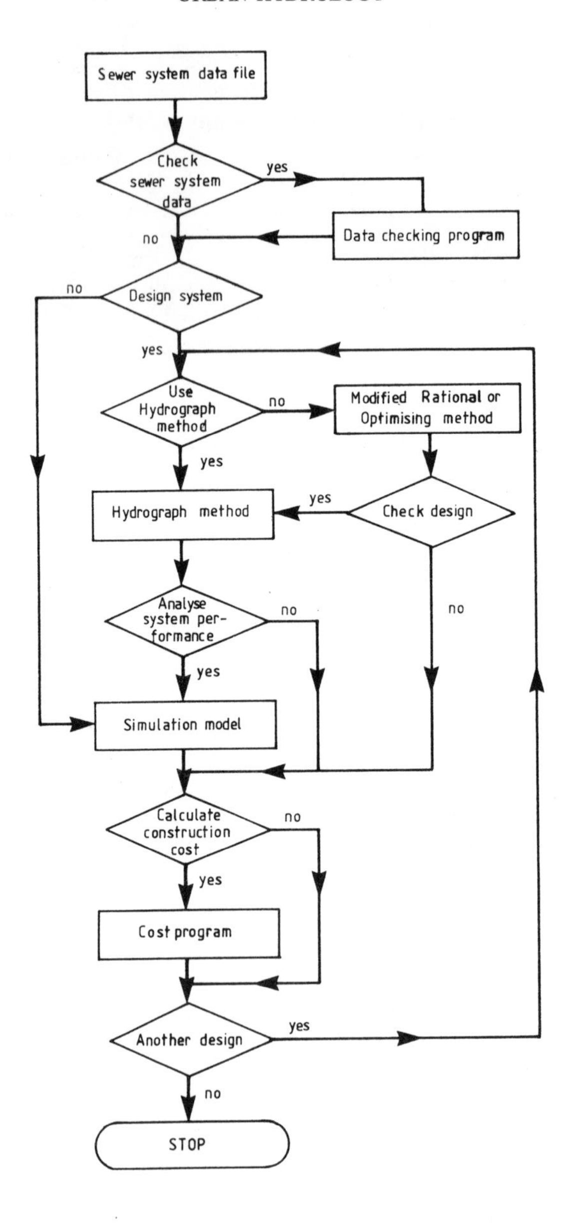


FIG. 9.8. Flowchart for the Wallingford Procedure (redrawn from National Water Council and Department of the Environment, 1981, by courtesy of Hydraulics Research Ltd, Wallingford, UK).

appropriate if the longitudinal profile of the sewer is constrained by the presence of other underground services.

If optimisation of pipe diameter, depth and gradient is not required, then the designer may use either the Modified Rational or the Hydrograph Methods. The former provides an estimate of peak discharge only, whereas the latter also produces the flood hydrograph. Both methods make allowances for the presence of certain types of ancillary structures.

For the analysis of an existing system, the Modified Rational or the Hydrograph Methods or the Simulation Program may be used (see Fig. 9.8). The Modified Rational Method is, as before, limited to the estimation of peak flow rates. The Simulation Program incorporates the same algorithm for simulating the above-ground phase of runoff as the Hydrograph Method. The pipe-routing technique is also the same until surcharging begins, and so both of these methods should yield similar results in non-surcharged pipe systems.

For both the design of new systems and the simulation of existing sewer networks, different methods may be more appropriate at different stages of an investigation. The Modified Rational Method may be applied for both design and analysis in order to provide an initial appreciation of catchment response. For a new sewerage system, the Optimising Method might then be employed to determine pipe sizes, depths and gradients, which subsequently can be checked using the Hydrograph Method. The latter approach can also be applied to check an existing system for surcharging. Finally, the Simulation Program both allows the performance of a proposed sewer network to be evaluated when subjected to rarer events than the selected design storm, and permits a more detailed examination of zones of surcharging in an existing pipe system.

Further details of both the Modified Rational and the Hydrograph Methods are summarised in the following sections.

### 9.5.1. The Wallingford Rational Method

In this method, eqn (9.1) is replaced by

$$Q_p = 2{\cdot}78 C_v C_R i A \qquad (9.17)$$

where $C_v$ is the volumetric runoff coefficient, $C_R$ is a routing coefficient which allows for non-linearity in the shape of the time–area diagram and variations in rainfall intensity within the time of concentration, and $Q_p$, $i$ and $A$ are as previously defined. If the total catchment area is being considered, the value of $C_v$ is computed from

$$C_v = \text{PR}/100 \qquad (9.18)$$

where PR, the percentage runoff, is given by

$$\text{PR} = 0{\cdot}829\text{IMP} + 25{\cdot}0\text{SOIL} + 0{\cdot}078\text{UCWI} - 20{\cdot}7 \qquad (9.19)$$

In eqn (9.19), IMP is the percentage impervious area of the catchment draining to the sewer, SOIL is a soil index which, for the United Kingdom, may be read from a map published by the Institute of Hydrology (1978), and UCWI is an antecedent wetness index which, for design purposes, is obtained from a relationship with the average annual rainfall. If impervious area alone is being considered,

$$C_v = \text{PR}/\text{IMP} \qquad (9.20)$$

For design purposes, a $C_R$ value of 1·3 has been recommended (National Water Council and Department of the Environment, 1981).

As before, the time of concentration is considered to consist of the sum of the time of flow (based upon full-bore pipe velocities) and a time of entry. The latter varies with both the design return period and the slope and size of the catchment area, ranging from 3–6 min for a 5-year event to 4–8 min for a one-year storm, the smaller values being applicable to the smaller, steeper catchments.

### 9.5.2. The Wallingford Hydrograph Method

This method was developed partly in response to criticism of the simplifications inherent in the TRRL Method with regard to:

(i) the representation of the above-ground phase of runoff by a time of entry;
(ii) the assumption of 100% runoff from the paved and no runoff from the pervious areas of a catchment;
(iii) storage allowances based solely on the pipe system with no attenuation attributed to above-ground storages; and
(iv) the assumption that the storm profile of a selected return period produces a peak discharge of the same return period.

Point (iv) in the above list is a general criticism that can be levelled at almost all flood estimation methods. In the Wallingford Hydrograph Method the relationship between the return period of the peak discharge and the return period of the causative design storm is maintained by the use of a stable set of design inputs. The latter have been

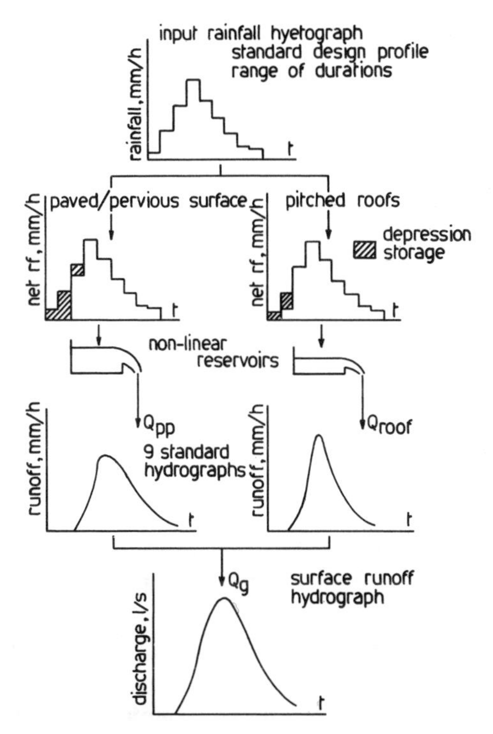


FIG. 9.9. The Wallingford Hydrograph Method: the above-ground phase of runoff (redrawn from National Water Council and Department of the Environment, 1981, by courtesy of Hydraulics Research Ltd, Wallingford, UK).

chosen by applying a technique described by Packman and Kidd (1980) involving the comparison of observed and computed probability distributions of peak flow rates, an outline of which has already been presented in Section 4.4. Points (i) to (iii) are countered by the separate modelling of the above-ground and the below-ground phases of runoff.

As shown in Fig. 9.9, the above-ground model in the Hydrograph Method consists of several components. For design purposes, the input to the model consists of a standard summer storm profile whose peakedness is exceeded by 50% of all such events (see Chapter 3). A duration of 15 min is assumed initially, and the computations are subsequently repeated for further values of 30, 60 and 120 min. In all cases, the total rainfall depth, $P$ (mm), is that corresponding to each duration and the return period of the peak discharge which is to be estimated.

Estimation of the losses on the subcatchments draining to each pipe length begins with the prediction of the percentage runoff, PR, from the whole catchment using eqn (9.19), which is then distributed between the paved surfaces, pitched roofs and pervious areas within each subcatchment. By observation, the percentage runoff from the impervious surfaces of small drainage areas was found to average 70%. If therefore the value of PR predicted from eqn (9.19) is less than 70% of the proportion of impervious surfaces within a subcatchment, the pervious areas are assumed not to contribute to storm runoff, so that

$$PR_{perv} = 0; \quad PR_{pav} = PR_{roof} = 100PR/IMP \tag{9.21}$$

where the subscripts pav, roof and perv refer to the paved, roof and pervious areas respectively. However, if PR exceeds 70% of the proportion of impervious area, the excess is distributed equally to all surfaces, giving

$$\begin{aligned} PR_{perv} &= PR - 0{\cdot}7IMP \\ PR_{pav} = PR_{roof} &= 70 + PR_{perv} \end{aligned} \tag{9.22}$$

Having obtained the appropriate percentage runoffs for each subcatchment, the distribution of effective rainfall is obtained from the storm profile by allowing for both an initial loss to depression storage, DS (mm), and a continuing loss by infiltration. For the paved and pervious areas,

$$DS_{perv} = DS_{pav} = 0{\cdot}71SLOPE^{-0{\cdot}48} \tag{9.23}$$

where SLOPE is the average overland slope (%) of the subcatchment. In practice, the necessity to take detailed measurements of slopes is avoided by allocating each subcatchment to one of three broad categories, as follows:

| *Description* | *Range in slope* | *Assumed value* |
|---|---|---|
| Mild | <2% | 1·25% |
| Medium | 2–3·5% | 2·75% |
| Steep | >3·5% | 4·0% |

For pitched roofs, a value of 0·4 mm is recommended for $DS_{roof}$.

Once depression storage has been subtracted from the beginning of the storm, the remaining loss is distributed uniformly throughout the rest of the duration by means of a reduced contributing area. Denoting the actual paved area within a subcatchment as $AREA_{pav}$, the contributing area, $AR_{pav}$, is given by:

$$AR_{pav} = AREA_{pav}[(PR_{pav}/100)\{P/(P - DS_{pav})\}] \tag{9.24}$$

Similar relationships are applicable for $AR_{perv}$ and $AR_{roof}$.

Next, the attenuation caused by surface storage is simulated by means of a non-linear reservoir, for which the storage volume, $S$, is related to the outflow discharge, $Q$, by the equation

$$S = KQ^{2/3} \tag{9.25}$$

where $K$ is the storage constant. Using data from a selection of catchments having both paved and pervious surfaces, the following prediction equation was obtained for $K$:

$$K = 0{\cdot}051\text{SLOPE}^{-0{\cdot}23}\text{PAPG}^{0{\cdot}23} \tag{9.26}$$

where PAPG is the average paved area per gulley. If the number of gulleys in each subcatchment is specified, PAPG may be computed directly. Otherwise, a characteristic value may be obtained by allocating each subcatchment to one of three broad categories, as follows:

| *Description* | *Range in area* | *Assumed value* |
|---|---|---|
| Small | $<200$ m$^2$ | 125 m$^2$ |
| Medium | 200–400 m$^2$ | 300 m$^2$ |
| Large | $>400$ m$^2$ | 600 m$^2$ |

The value of $K$ computed from eqn (9.26) is applied to the effective contributing area, $AR_{perv} + AR_{pav}$, of the paved and pervious surfaces together. For pitched roofs, a value of $K$ of 0·04 is recommended.

Even with such a simple model for attenuation as that given by eqn (9.25), the amount of computation can become excessive with even a modest number of subcatchments. The calculations are therefore simplified by the use of nine 'standard' subcatchments, defined by three values each of SLOPE and PAPG as given in the above tables. The runoff hydrographs (mm/h) from each of the nine standard subcatchments are computed initially, and every actual subcatchment is then represented by one of the nine. A roof hydrograph may also be synthesised, if required. The gulley hydrograph is obtained by adding the roof hydrograph with its ordinates multiplied by $AR_{roof}$ to the appropriate standard hydrograph with its ordinates multiplied by $AR_{pav} + AR_{perv}$:

$$Q_g = [Q_{roof}AR_{roof} + Q_{pp}(AR_{pav} + AR_{perv})]/3600 \qquad (9.27)$$

where $Q_{roof}$ and $Q_{pp}$ are the ordinates of the roof and standard hydrographs (mm/h) respectively, and $Q_g$ are the ordinates of the gulley hydrograph (litres/s). The hydrographs obtained from eqn (9.27) for each subcatchment are then routed through the sewer network, pipe by pipe, using the Muskingum–Cunge Method (see Chapter 7). The calculations are carried out for all four standard durations of storm profile, and the largest computed discharges are taken as the design flows for each pipe length.

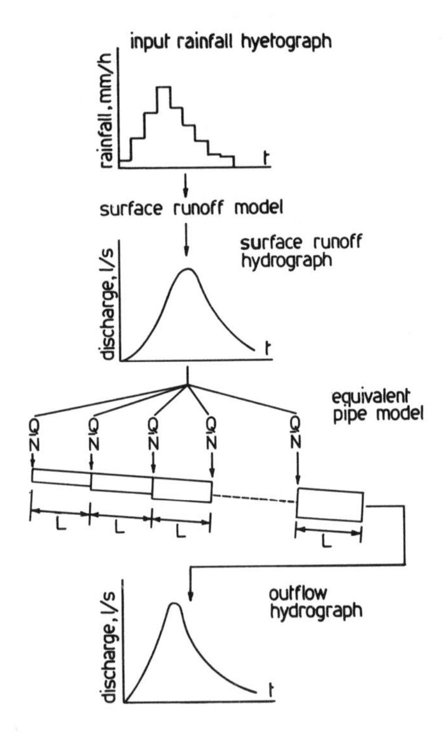


FIG. 9.10. The Wallingford Hydrograph Method: the Sewered Sub-area Model (redrawn from National Water Council and Department of the Environment, 1981, by courtesy of Hydraulics Research Ltd, Wallingford, UK).

Where insufficient data are available to permit the modelling of both the above-ground and the below-ground phases of runoff for every subcatchment and pipe length, or where the costs of data collection for a large drainage area would be prohibitive, a simplified sub-area model is available. In this model, the method of computing the gulley hydrographs is applied to sub-areas of up to 60 ha instead of each pipe length. As shown schematically in Fig. 9.10, the computer sub-area hydrograph is then divided into *N* equal parts and distributed equally to the *N* segments of an 'equivalent pipe'. The latter consists of a tapered system of pipes in series, each of which has the same length and slope. The number of segments, *N*, depends upon the time of flow within the equivalent pipe. The model requires as input data the total length of the major pipe run in the sub-area, the average pipe slope, and the diameter and slope of the outfall pipe. Where no details of the outflow pipe are available, as in a design application, its dimensions must be estimated using the Modified Rational Method. Using this Sewered Sub-area Model, substantial savings on input data are possible, with networks of the order of 100 pipes being reduced to only four equivalent pipes. As before, routing of flows through the equivalent pipes is carried out using the Muskingum–Cunge Method.

## REFERENCES


AITKEN, A. P. (1973) Hydrologic investigation and design in urban areas—a review. Austr. Wat. Resour. Council, Tech. Paper No. 5, 79 pp.

APPLEBY, F. V. (1937) Impervious factors. *Proc. Instn. Munic. Co. Engrs.*, **63,** 1077–1100.

ARDIS, C. V., DUEKER, K. J. and LENZ, A. T. (1969) Storm drainage practices of thirty-two cities. *Proc. Am. Soc. Civ. Engrs., J. Hydraul. Div.*, **95**(HYl), 383–408.

BETTESS, R., PITFIELD, R. A. and PRICE, R. K. (1978) A surcharging model for storm sewer systems. In Helliwell, P. R. (ed.), *Urban storm drainage*, Proc. Int. Conf., Southampton (Pentech Press, London) pp. 306–16.

CHEN, C. W. and SHUBINSKI, R. P. (1971) Computer simulation of urban storm water runoff. *Proc. Am. Soc. Civ. Engrs., J. Hydraul. Div.*, **97**(HY2), 289–301.

CHOW, V. T. (1964) Runoff. Section 14 of Chow, V. T. (ed.), *Handbook of applied hydrology* (McGraw-Hill, New York) 54 pp.

COLEMAN, G. S. and JOHNSON, A. (1932) Rainfall runoff. *Proc. Instn. Munic. Co. Engrs.*, **58,** 1403–15.

COLYER, P. J. (1977) Performance of storm drainage simulation models. *Proc. Instn. Civ. Engrs., Part 2*, **63,** 293–309.

COLYER, P. J. and PETHICK, R. W. (1976) Storm drainage design methods: a literature review. Hydraul. Res. Station, Wallingford, Rept. No. INT 154, 85 pp.

COOK, L. A. and LOCKWOOD, B. (1977) The investigation of sewer networks by computer. *Proc. Instn. Civ. Engrs., Part 2*, **63,** 481–94.

CUNGE, J. A. (1974) Evaluation problem of storm water routing mathematical models. *Wat. Res.*, **8,** 1083–7.

DOOGE, J. C. I. (1957) The Rational Method for estimating flood peaks—Irish contributions to the technique. *Engineering*, **184,** 311–13, 374–7.

ESCRITT, L. B. (1965) *Sewerage and sewage disposal*, 3rd Edn (CR Books/Applied Science Publishers, London) 488 pp.

ESCRITT, L. B. and YOUNG, A. J. M. (1963) Economic surface water sewerage: a suggested standard of practice. *J. Instn. Pub. Health Engrs.*, **62,** 333–85.

GLUCK, W. R. and MCCUEN, R. H. (1975) Estimating land use characteristics for hydrologic models. *Wat. Resour. Res.*, **11,** 177–9.

HALL, M. J. and HOCKIN, D. L. (1980) Guide to the design of storage ponds for flood control in partly urbanised catchment areas. Construction Industry Research and Information Association, Tech. Note 100, 103 pp.

HART, C. A. (1933) Rainfall and runoff (correspondence). *Proc. Instn. Munic. Co. Engrs.*, **59,** 978–80.

HAWKEN, W. H. (1921) An analysis of maximum runoff and rainfall intensity. *Trans. Instn. Engrs. Austr.*, **2,** 193–215.

HEEPS, D. P. and MEIN, R. G. (1973) An independent evaluation of three urban stormwater models. Monash Univ., Civ. Engng. Res. Rept. No. 4/1973, 92 pp.

HEEPS, D. P. and MEIN, R. G. (1974) Independent comparison of three urban runoff models. *Proc. Am. Soc. Civ. Engrs., J. Hydraul. Div.*, **100**(HY7), 995–1009.

HYDRAULIC DESIGN OF STORM SEWERS WORKING PARTY (1976) A review of progress, March 1974–June 1975. Department of the Environment and National Water Council, Standing Technical Committee Reports No. 1, 26 pp.

INSTITUTE OF HYDROLOGY (1978) A revised version of the winter rain acceptance potential map. Institute of Hydrology, Wallingford, Flood Studies Suppl. Rept. No. 7, 1 p.

IZZARD, C. F. (1946) Hydraulics of runoff from developed surfaces. *Proc. Highway Res. Bd.*, **26,** 129–46.

JUDSON, C. C. (1933) Runoff calculations, a new method. *Proc. Instn. Munic. Co. Engrs.*, **59,** 861–7.

KIDD, C. H. R. (1978) A calibrated model for the simulation of the inlet hydrograph for fully sewered catchments. In Helliwell, P. R. (ed.), *Urban storm drainage*, Proc. Int. Conf., Southampton (Pentech Press, London) pp. 172–86.

KIDD, C. H. R. and HELLIWELL, P. R. (1977) Simulation of the inlet hydrograph for urban catchments. *J. Hydrol.*, **35,** 159–72.

KUICHLING, E. (1889) The relation between the rainfall and the discharge of sewers in populous districts. *Trans. Am. Soc. Civ. Engrs.*, **20,** 1–60.

LAGER, J. A., SHUBINSKI, R. P. and RUSSELL, L. W. (1971) Development of a simulation model for stormwater management. *J. Wat. Poll. Contr. Fed.*, **43,** 2424–35.

LLOYD-DAVIES, D. E. (1906) The elimination of stormwater from sewerage systems. *Min. Proc. Instn. Civ. Engrs.*, **164,** 41–67.

MACLAREN, J. F., LTD (1975) Review of Canadian design practice and comparison of urban hydrologic models. Ontario Ministry of the Environment, Res. Progr. for the Abatement of Munic. Poll., Res. Rept. No. 26, 224 pp.

MARSALEK, J., DICK, T. M., WISNER, P. E. and CLARK, W. G. (1975) Comparative evaluation of three urban runoff models. *Wat. Resour. Bull.*, **11,** 306–28.

MARTIN, C. and KING, D. (1978) Improvements in the TRRL hydrograph program. In Helliwell, P. R. (ed.), *Urban storm drainage*, Proc. Int. Conf., Southampton (Pentech Press, London) pp. 207–25.

MAYS, L. W. and YEN, B. C. (1975) Optimal cost design of branched sewer systems. *Wat. Resour. Res.*, **11,** 37–47.

MEEK, J. B. L. (1928) Sewerage, with special reference to runoff. *Instn. Civ. Engrs., Conf. Rept. Disc.*, pp. 162–74.

MULVANEY, T. J. (1851) On the use of self-registering rain and flood gauges in making observations on the relation of rainfall and of flood discharges in a given catchment. *Trans. Instn. Civ. Engrs. Ireland*, **4**(2), 18–31.

MUZIK, I. (1974) Laboratory experiments with surface runoff. *Proc. Am. Soc. Civ. Engrs., J. Hydraul. Div.*, **100**(HY4), 501–16.

NASH, J. E. (1958) Determining runoff from rainfall. *Proc. Instn. Civ. Engrs.*, **10,** 163–84.

NATIONAL WATER COUNCIL and DEPARTMENT OF THE ENVIRONMENT (1981) *Design and Analysis of Urban Storm Drainage: The Wallingford Procedure*, Vol. 1, *Principles, methods and practice*; Vol. 2, *Program users' guide*; Vol. 3, *Maps*; Vol. 4, *Modified Rational Method* (National Water Council, London).

NORRIS, W. H. (1946) Estimation of runoff from impervious surfaces. *Proc. Instn. Munic. Co. Engrs.*, **72,** 425–38.

ORMSBY, M. T. M. (1933) Rainfall and runoff calculations. *Proc. Instn. Munic. Co. Engrs.*, **59,** 889–94.

PACKMAN, J. C. and KIDD, C. H. R. (1980) A logical approach to the design storm concept. *Wat. Resour. Res.*, **16,** 994–1000.

PAPADAKIS, C. N. and PREUL, H. C. (1972) University of Cincinnati urban runoff model. *Proc. Am. Soc. Civ. Engrs., J. Hydraul. Div.*, **98**(HY10), 1789–1804.

PAPADAKIS, C. N. and PREUL, H. C. (1973) Testing of methods for determination of urban runoff. *Proc. Am. Soc. Civ. Engrs., J. Hydraul. Div.*, **99**(HY9), 1319–35.

PRICE, R. K. (1978) Design of storm sewers for minimum construction cost. In Helliwell, P. R. (ed.), *Urban storm drainage*, Proc. Int. Conf., Southampton (Pentech Press, London) pp. 636–47.

PRICE, R. K. and KIDD, C. H. R. (1978) A design and simulation method for storm sewers. In Helliwell, P. R. (ed.), *Urban storm drainage*, Proc. Int. Conf., Southampton (Pentech Press, London) pp. 327–37.

Reid, J. (1927) The estimation of storm water discharge. *J. Instn. Munic. Co. Engrs.*, **53,** 997–1027.

Riley, D. W. (1932) Notes on calculating the flow of surface water in sewers. *Proc. Instn. Munic. Co. Engrs.*, **58,** 1483–94.

Road Research Laboratory (1963) A guide for engineers to the design of storm sewer systems. Road Note No. 35 (HMSO, London) 20 pp.

Roseveare, L. (1930) Runoff as affecting the flow in sewers. *Proc. Instn. Munic. Co. Engrs.*, **56,** 1177–97.

Ross, C. N. (1921) The calculation of flood discharges by the use of a time contour plan. *Trans. Instn. Engrs. Austr.*, **2,** 85–92.

Sarginson, E. J. and Bourne, D. E. (1969) The analysis of urban rainfall runoff and discharge. *J. Instn. Munic. Engrs.*, **96,** 81–5.

Schaake, J. C. (1969) A summary of the Hopkins storm drainage research project: its objectives, its accomplishments and its relation to future problems in urban hydrology. In *Proc. 1st Int. Seminar for Hydrol. Professors,* Vol. II, *Specialised hydrologic subjects* (Dept. of Civil Engineering, Univ. of Illinois at Urbana–Champaign) pp. 638–65.

Schaake, J. C., Geyer, J. C. and Knapp, J. W. (1967) Experimental examination of the Rational Method. *Proc. Am. Soc. Civ. Engrs., J. Hydraul. Div.*, **93**(HY6), 353–70.

Shen, H. W. and Li, R. M. (1973) Rainfall effect on sheet flow over smooth surface. *Proc. Am. Soc. Civ. Engrs., J. Hydraul. Div.*, **99**(HY5), 771–92.

Swinnerton, C. J., Hall, M. J. and O'Donnell, T. (1973) Conceptual model design for motorway stormwater drainage. *Civ. Engng. Pub. Wks. Rev.*, **68,** 123–9, 132.

Terstriep, M. L. and Stall, J. B. (1969) Urban runoff by Road Research Laboratory method. *Proc. Am. Soc. Civ. Engrs., J. Hydraul. Div.*, **95**(HY6), 1809–34.

Terstriep, M. L. and Stall, J. B. (1974) Illinois urban drainage area simulator, ILLUDAS. Illinois State Wat. Survey, Bull. No. 58, 90 pp.

Tholin, A. L. and Keifer, C. J. (1960) Hydrology of urban runoff. *Trans. Am. Soc. Civ. Engrs.*, **125,** 1308–79.

Thompson, J. L. and Lupton, A. R. R. (1978) A method of assessment of piped drainage systems taking account of surcharge and overground flooding. In Helliwell, P. R. (ed.), *Urban storm drainage,* Proc. Int. Conf., Southampton (Pentech Press, London) pp. 226–42.

Torno, H. C. (1975) A model for assessing impact of storm-water runoff and combined sewer overflows and evaluating pollution abatement alternatives. *Wat. Res.*, **9,** 813–15.

Transport and Road Research Laboratory (1976) A guide for engineers to the design of storm sewer systems. Road Note No. 35, 2nd Edn (HMSO, London) 30 pp.

Viessman, W. (1966) The hydrology of small impervious areas. *Wat. Resour. Res.*, **2,** 405–12.

Viessman, W. (1968) Run-off estimation for very small drainage areas. *Wat. Resour. Res.*, **4,** 87–93.

Viessman, W., Keating, W. R. and Srinivasa, K. N. (1970) Urban storm runoff relations. *Wat. Resour. Res.*, **6,** 275–9.

WATER POLLUTION CONTROL FEDERATION and AMERICAN SOCIETY OF CIVIL ENGINEERS (1969) *Design and construction of sanitary and storm sewers.* WPCF Manual of Practice No. 9 (ASCE Manuals and Reports on Engng. Practice No. 37, New York) 332 pp.

WATKINS, L. H. (1956) Rainfall and runoff, an investigation at Harlow New Town. *J. Instn. Munic. Engrs.*, **82,** 305–16.

WATKINS, L. H. (1962) The design of urban sewer systems: research into the relation between the rate of rainfall and the rate of flow in sewers. Road Res. Tech. Paper No. 55 (HMSO, London) 96 pp.

WATKINS, L. H. (1963) Research on surface-water drainage. *Proc. Instn. Civ. Engrs.*, **24,** 305–30.

WATKINS, L. H. (1976) The TRRL hydrograph method of urban sewer design adapted for tropical conditions. *Proc. Instn. Civ. Engrs., Part* 2, **61,** 539–66.

WATKINS, L. H. and FIDDES, D. (1978) The design of surface water sewer systems in the tropics. In Helliwell, P. R. (ed.), *Urban storm drainage*, Proc. Int. Conf., Southampton (Pentech Press, London) pp. 243–55.

WATKINS, L. H. and YOUNG, C. P. (1965) Developments in urban hydrology in Great Britain. Road Research Laboratory, Lab. Note No. LN/885/LHW.CPY, 18 pp.

WATT, W. E. and KIDD, C. H. R. (1975) QUURM—a realistic urban runoff model. *J. Hydrol.*, **27,** 225–35.

WILLEKE, G. E. (1966) Time in urban hydrology. *Proc. Am. Soc. Civ. Engrs., J. Hydraul. Div.*, **92**(HY1), 13–29.

WILLIAMS, G. R. (1950) Hydrology. Ch. IV of Rouse, H. (ed.), *Engineering hydraulics* (Wiley, New York) pp. 229–320.

WOO, D. C. and BRATER, E. F. (1962) Spatially varied flow from controlled rainfall. *Proc. Am. Soc. Civ. Engrs., J. Hydraul. Div.*, **88**(HY6), 31–56.

YEN, B. C. and SEVUK, A. S. (1975) Design of storm sewer networks. *Proc. Am. Soc. Civ. Engrs., J. Environ. Engng. Div.*, **101**(EE4), 535–53.

YEVJEVICH, V. and BARNES, A. H. (1970) Flood routing through storm drains, Part IV: Numerical computer methods of solution. Colorado State Univ., Hydrol. Paper No. 46, 47 pp.

YOON, N. Y. and WENZEL, H. G. (1971) Mechanics of sheet flow under simulated rainfall. *Proc. Am. Soc. Civ. Engrs., J. Hydraul. Div.*, **97**(HY9), 1367–86.

YOUNG, C. P. (1973) Urban drainage in the United Kingdom. In Construction Industry Research and Information Association, *Proc. Res. Colloquium on Rainfall, Runoff and Surface Water Drainage of Urban Catchments*, Proc. Paper 1.

YU, Y. S. and MCNOWN, J. S. (1964) Runoff from impervious surfaces. *J. Hydraul. Res.*, **2,** 3–24.

ZAGHLOUL, N. A. (1977) Evaluation of storm water routing models. *Proc. Instn. Civ. Engrs., Part 2*, **63,** 925–33.

# Part IV

# WATER QUALITY PROBLEMS OF URBAN AREAS

# 10

# Water Quality Changes Caused by Urbanisation

## 10.1. INTRODUCTION

Even with the enhanced capabilities of modern methods of sampling and analysis, the complete evaluation of the quality of a body of water is a mammoth, if not impossible, task. In its natural or ambient state, water quality is governed by a complex of geological, hydrological and topographical controls, overlain by a multiplicity of geochemical and biochemical processes. Objective assessment of the changes in water quality brought about by human activities in urban areas presupposes a knowledge of this ambient state, yet in a large receiving water body such conditions may well include a fully mixed effluent from an upstream discharge point. In attempting to specify maximum (or minimum) allowable figures for concentrations of chemical constituents or physical properties in receiving waters or effluent discharges, water quality standards may perhaps be seen as upper (or lower) limits to what informed opinion regards as 'natural' conditions.

That few records of water quality were maintained prior to 1960 provides a reflection of the relatively limited extent of public and professional concern at that time compared with the present day. The casual observer might therefore be forgiven for assuming that scientific interest in the quality of water from urban areas is comparatively recent. However, as Waller (1972) and others have pointed out, the composition of urban runoff featured prominently in the debate on the relative merits of combined and separate sewerage systems in the later nineteenth century. Unfortunately, this interest was not sustained into the early twentieth century, possibly because of a temporary improvement in water quality as the use of horses declined with the advent of the internal combustion engine. In any case, such considerations were relatively minor at a time when the majority of towns and cities were discharging their domestic and industrial wastes into the nearest watercourse without treatment.

The Royal Commission on Sewage Disposal, which was set up by the British Government in 1889, is regarded by many as laying the foundations of modern sewage treatment and effluent control (Sidwick, 1977). Of the ten reports that the Royal Commission produced during its seventeen-year existence, two have had a lasting effect, at least in the British Isles. The third report recommended a national and regional organisation for sewage treatment and control that was finally adopted with the creation of Regional Water Authorities under the Water Act 1973. The eighth report established the so-called Royal Commission effluent standards, which having been given general legal status by the Rivers (Prevention of Pollution) Act 1951, are still familiar in present-day practice. Such has been the influence of the Royal Commission and its work that even the period since the Second World War has been characterised by Sidwick (1977) as one of continuing refinement rather than original development. This general lack of innovation was attributed by Sidwick to the capability of conventional treatment processes to achieve the increasingly stringent effluent standards demanded by authorities and general public alike.

The raising of effluent standards has not only stimulated research and development, but has also drawn attention to sources of pollution other than those under direct local government control. For convenience, the origins of contaminants within a given catchment area are frequently classified as either point or non-point sources. The former are generally regarded as the discharges to watercourses that are concentrated at a single outfall, such as a stormwater drain, the overflow from a combined sewerage system or the effluent from a sewage treatment plant. In contrast to the discrete nature of a point source, the origins of non-point source pollution are diffused over a large area, and are exemplified by substances such as sediment and agricultural chemicals. The principal non-point source of pollution is therefore the surface water runoff from the catchment. Consequently, any water quality problems which arise from non-point sources are intimately related to the dominant land use within the drainage area.

Given these definitions, the evaluation of water quality may be largely regarded as a problem in distinguishing between the contributions of pollutants from both point and non-point sources. For planning purposes, Whipple *et al.* (1974) recommended that a further distinction should be drawn between recorded and unrecorded sources, i.e. those which may or may not be under the direct control of the local authorities. An extra dimension is added to the problem by the need to specify which water quality constituents are to be monitored and the methods of sampling that are to be employed. This topic is therefore discussed in further detail in Section 10.2 before considering the available information on the temporal and spatial variations in urban water quality in Section 10.3.

## 10.2. THE CHARACTERISATION AND MONITORING OF WATER QUALITY

As James (1976) has noted, work by the Royal Commission on Sewage Disposal among others established the relationship between the deoxygenation of watercourses and the bacterial decomposition of organic matter, so that organic content has long been recognised as the principal criterion of water quality. The amount of dissolved oxygen (DO) and the temperature are frequently employed as a measure of the general health of a water body. However, since the organic matter in natural waters can be of such diverse composition, assessment is usually made indirectly using the biochemical oxygen demand (BOD) test, in which the amount of oxygen consumed by micro-organisms in metabolism is measured under standardised conditions. This test is generally carried out over a standard period, 5 and 28 days being the most common durations. The BOD test gives the amount of organic matter that is bacterially degradable. A similar chemical oxygen demand (COD) test is available for the assessment of total organic content.

A second widely applied criterion of water quality is the content of solids including bacteria. A distinction is drawn between dissolved solids (DS) and suspended solids (SS), with the former being distinguished by their ability to pass through a 45 $\mu$m filter. Dissolved solids only exert a biological influence at high levels that are more characteristic of marine environments. Moreover, dissolved solids do not change the appearance of a water body in the same manner as suspended solids. Owing to their effect upon turbidity and colour and, in the case of domestic and industrial wastes, the oxygen demand that they may exert, more attention is devoted to SS than to DS.

The importance of suspended sediments is two-fold. Firstly, high levels of SS can alter the structure and productivity of plant, invertebrate and vertebrate communities (see, for example, Barton, 1977). As noted by Karr and Schlosser (1978), in order to cause mortality in adult fish, SS concentrations must exceed 20 000 mg/litre. In addition, settling sediments have an indirect effect by covering essential spawning grounds. Secondly, SS act as sorbents for heavy metals, such as lead and cadmium (Angino *et al.*, 1972; Bryan, 1974; Ellis, 1976, 1977), mercury and chlorinated organics (Murphy and Carleo, 1978) and hydrocarbons (Hunter *et al.*, 1979). The potential concentrations in the benthal sludges of an urban watercourse are therefore high. More importantly, suspended solids provide the transporting medium for these constituents.

A third important criterion of water quality is the content of nutrients. Owing to their importance in the growth of algae and therefore in the control of eutrophication, particular attention is now paid to the inorganic salts of nitrogen and phosphorus, which are major constituents of fertilisers and detergents. Nutrient loadings are reported in a variety of forms. Total Kjeldahl nitrogen (TKN) is the sum of organic nitrogen and ammonia expressed as equivalent nitrogen; total nitrogen (TN) is the sum of TKN, nitrate and nitrite nitrogen as equivalent N. Total phosphorus (TP) and total phosphate as equivalent P (including the *meta*-phosphate and *ortho*-phosphate forms) are also commonly quoted.

Of the different forms of N, particular concern is often expressed about nitrate, owing to its association with infantile methaemoglobinaemia. The World Health Organisation European Standard for Drinking Water recommends that concentrations of nitrate as equivalent N should not exceed 11·3 mg/litre. However, Marsh (1980) has indicated that nitrate levels in English rivers have increased by up to 70% during the last 20 years, such that mean concentrations exceeding this limit are now observable at some sites in the Midlands and south-east of the country. In certain rivers, nitrate concentrations exhibit a seasonality, with levels remaining well above the mean for extended periods during the winter months.

With the increase in the volume of industrial wastewater discharges in many countries, the presence of toxins provides another criterion of water quality. Of the common aquatic toxins listed by James (1976), heavy metals, such as lead, zinc, cadmium, copper, nickel and mercury, and pesticides and herbicides, such as DDT, are of particular concern. Of the heavy metals, lead has attracted special attention because of its use in motor vehicle fuels. According to Solomon *et al.* (1977), lead is emitted from vehicle exhausts at a rate of between 0·06 and 0·02 g/km. Roadside dust may contain up to 24 g Pb/$m^2$ of street surface. Its impact upon water quality is largely dependent upon the hydrological response of the drainage system, the background concentration in the receiving waters and the chemical forms of lead in both the runoff and the receiving waters (see Laxen and Harrison, 1977).

In listing water quality criteria, the physico-chemical parameters such as temperature, pH, turbidity and electrical conductivity should not be forgotten. Temperature is especially important because of its influence on the physiological processes of aquatic animals. The capacity of water to hold oxygen decreases as the temperature increases. Moreover, the rate at which nutrients attached to solid particles are converted to readily available soluble forms increases as the temperature is raised. According to Karr and Schlosser (1978), removal of streamside vegetation can increase water temperatures by 6–9°C, which can be sufficient to eliminate species with low-temperature optima.

The above list by no means includes all the criteria that can be employed to characterise water quality. For example, microbiological quality is particularly relevant because of the dangers to public health of high levels of bacteria and pathogenic micro-organisms, but studies in relation to urban runoff, such as that of Qureshi and Dutka (1979), have as yet been relatively infrequent. In more general terms, the marked improvement in the sensitivity of analytical techniques during the last decade has allowed the presence of contaminants in water to be detected and quantified at very low concentrations. Nevertheless, difficulties continue to be encountered in analysing organic compounds, an extensive variety of which are produced by various industries. Many of these compounds must be discharged to the aquatic environment in wastewater. Since no two methods of analysis tend to recover the same set of compounds from a given source, Janardan and Schaeffer (1981) have provided a method for estimating the number of ‘undiscovered’ compounds to be expected in any sampling programme.

Current knowledge of the abundance of potential contaminants in a water body greatly assists in identifying sampling requirements. However, much work remains to be done in improving sampling techniques. Reliable information on the temporal variations in water quality can only be obtained if representative samples can be taken at regular intervals throughout the duration of the storm hydrograph. Manual methods, involving 'grab' sampling by field personnel, tend to be unreliable, especially when major events occur outside the normal working day, and Daniel *et al.* (1978) have argued strongly for the use of automatic sampling equipment. However economical in terms of installation and running costs such devices may be, the costs associated with the subsequent analysis should not be underestimated. The latter depend upon both the nature and the number of constituents. Heavy metals and pesticides are particularly expensive to analyse, and Daniel *et al.* (1978) warn against the tendency to include water quality parameters that are not immediately relevant to the purposes of the monitoring programme.

In contrast, Whipple and McIntosh (1979) have suggested that for many purposes information on the general condition of urban watercourses can be collected without undertaking costly measurements and analysis. Those authors outline a stream surveillance method that concentrates upon the biological health of streams, using indicators such as macro-invertebrates that respond quickly to changes in habitat. Additional attention may be paid to such factors as channel erosion, obstructions to flow and channel infilling as well as odours, which might indicate the presence of sewage overflows, and turbidity, which could be associated with increased SS loadings resulting from construction activity.

Reference has already been made to the importance of sediment, both for the removal of heavy metals and organic materials from solution and as a medium for their transportation. However, the movement of sediment in a natural watercourse is an extremely complex process. The different components of sediment transport are defined in Fig. 10.1. The total load transported consists of bed load, which moves by rolling or saltation and is in almost continuous contact with the bed, and suspended load, which is maintained in suspension by turbulence away from the stream bed. At its point of origin, the total load comprises the bed material load (including the suspended portion) and the wash load, which is composed of particle sizes smaller than those found in the bed material that are virtually permanently in suspension. The concentration of suspended sediment varies both with depth and with distance from the banks, and also fluctuates with time at any given point in the cross-section. Multi-point sampling procedures are therefore required in order to obtain a reliable estimate of suspended sediment (i.e. suspended solids) loadings (see, for example, Graf, 1971, Ch. 13; Vanoni, 1975, Ch. III), but such techniques are

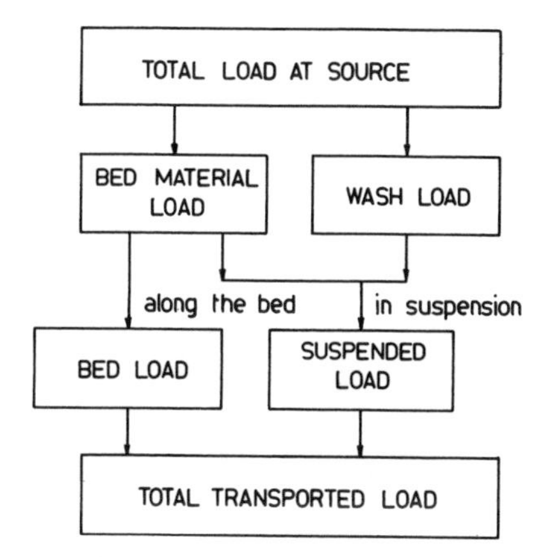


FIG. 10.1. Definition of the components of sediment transport.

seldom applied in water quality studies. Much remains to be learned about appropriate types of sampling devices and the most efficient sampling procedures.

## 10.3. SPATIAL AND TEMPORAL VARIATIONS IN WATER QUALITY

The factors which influence the quality of water within an urban area are broadly indicated in Fig. 10.2. The rain falling upon the urban area is itself contaminated, and its constituents may vary widely from place to place. Lewis and Grant (1978) have suggested that, from an analytical viewpoint, total dry and wet precipitation can be considered to consist of three fractions:

(1) dissolved materials in aqueous precipitation;
(2) the water-soluble component of either dry or wet precipitation; and
(3) the water-insoluble component of either dry or wet precipitation.

According to those authors, (3) above is typically ignored, and the types of samplers generally in use lead to a highly variable mixture of all three fractions. What is generally termed ‘bulk precipitation’ (see Whitehead and Feth, 1964) includes (1) and (2) but not (3). Nevertheless, Lewis and Grant (1978) have demonstrated that, with a suitable

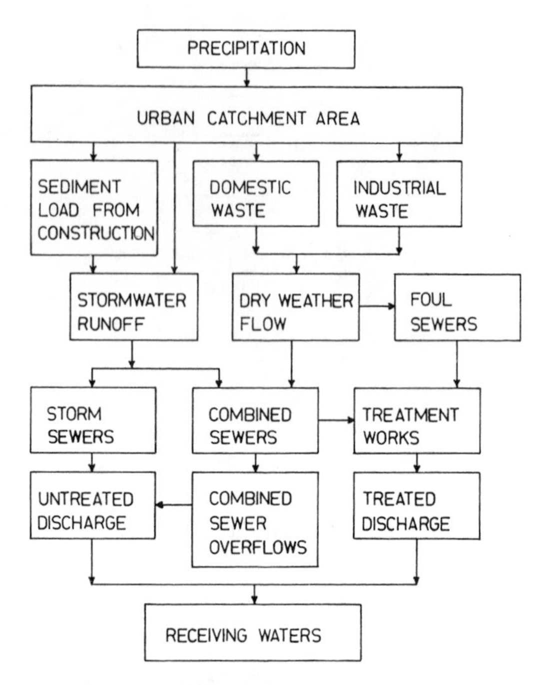


FIG. 10.2. Water quality changes in urban areas.

design of collector, meaningful chemical data for a small to moderate size of catchment can be obtained from a single sampling point, despite the high natural variability of rainfall.

Even with the limitations of the available sampling devices, several authors, including Barkdoll *et al.* (1977) and Betson (1978), have been able to demonstrate that atmospheric sources can account for a high proportion of many streamflow contaminants, although considerable variations in loading can be observed between different catchments. A particular characteristic of rainfall at urban sites appears to be a substantial COD and nutrient content (see Characklis *et al.*, 1979). Since a large proportion of the pollutants in street dust occur within the range of particle sizes that are only partially removed by street sweeping but are easily resuspended by winds (see Sartor *et al.*, 1974), constituents such as heavy metals can frequently be detected in undeveloped areas, as shown by Huff (1976), Burt and Day (1977) and Dethier (1979).

Climate also plays its part in determining the magnitude and seasonality in precipitation loadings. For example, Lewis (1981) has described a study in Venezuela in which 15% of the total annual loading was flushed out in a 2-week period at the beginning of the wet season. Such occurrences can lead to the production of acid rain, i.e. precipitation with a pH of less than 5·7, by the removal of oxides of nitrogen. Unfortunately, little attention has yet been paid to the water quality of urban catchments in the tropics, and so the significance of this effect with respect to runoff constituents has still to be evaluated.

The more marked changes in water quality occur within the land phase of the hydrological cycle (see Fig. 10.2). Stormwater runoff may contain a wide variety of pollutants, ranging from organic wastes to nutrients, bacteria, suspended solids and heavy metals. Surface runoff from streets typically contributes further contaminants, which may include garden wastes, fertilisers, pesticides, animal wastes, oils and detergents, tyre and vehicle exhaust residues, de-icing chemicals and miscellaneous dirt and street litter. A major source of water quality degradation in many developing urban areas is the increased sediment load derived from construction activities. In addition, waste loadings are contributed from both domestic and industrial sources, which together provide the 'dry weather flow' of the urban area.

The path of the pollutants carried by both the dry weather flow and the stormwater runoff, and their ultimate effect upon the receiving water body, depend largely upon the drainage system and its characteristics. In older towns and cities the sewerage is invariably constructed on the combined principle, with both foul sewage and stormwater runoff contained within the same pipe system. During dry weather flow the sewage can be treated before discharge into the receiving water body, but following any significant storm event the runoff volume is often too large to be handled at the local treatment works, and a large portion may be spilled untreated into the nearest watercourse via stormwater overflows. The latter are invariably constructed to operate when flows in the pipe system exceed six times the dry weather flow.

The concept of separate sewerage systems, in which foul sewage and stormwater are served by independent pipe networks, was introduced primarily to avoid the nuisance created by stormwater overflows on combined systems. The increased cost of duplicating pipe systems and house connections was thought to be offset by the savings which accrued from the elimination of storm tanks at the sewage works.

Moreover, the dry weather flow contains less grit and is subject to smaller variations in discharge. The stormwater is directed into the nearest watercourse without treatment.

In effect, the adoption of separate sewerage systems, as in the case of the New Town developments in the United Kingdom from 1950 onwards, was based upon the assumption that the composition of stormwater presents less of a pollution hazard than the overflow from a combined system. The revival of interest in this controversy, which appears to date from the study by Wilkinson (1956) on the quality of urban runoff from a typical British post-war housing estate, has also drawn attention to the more general problem of non-point source pollution and its management. More than 20 years later, the advantages and disadvantages of separate sewerage systems were again rehearsed by the Scottish Development Department (1977), and the impression remains that their cost implications are not always fully investigated at the design stage, particularly in the light of modern water pollution legislation. Indeed, for the type of conditions encountered in The Netherlands, where pipe gradients are flat and the dry weather flow is pumped to the treatment works, Wiggers and Bakker (1978) have concluded that combined systems are attractive from both the pollution control and the cost viewpoints.

The study by Wilkinson (1956) was specifically directed towards the composition of the stormwater runoff from a 247 ha housing estate with separate surface water sewers in the south-east of England. Using the results from some 700 samples taken over a 12-month period, Wilkinson was able to demonstrate that the loadings of BOD and SS in the stormwater runoff would be about 40 and 650% respectively of those that would be expected if both the runoff and the dry weather flow from the estate were treated to Royal Commission standards (20 mg/litre BOD and 30 mg/litre SS). Weibel *et al.* (1964), who carried out a similar study on an 11 ha urban residential and commercial area with separate sewers in Cincinnati, also showed that urban runoff compared unfavourably with domestic sewage with regard to annual loadings of SS.

Following the work of Wilkinson (1956) and Weibel *et al.* (1964), little further information on urban stormwater quality has become available until within the last 5–10 years. This renewed activity can be attributed to both the continuing improvement in analytical techniques and the reliability of automatic sampling equipment, and the growing public interest in the protection of the environment as exemplified by water pollution legislation on both sides of the Atlantic. This more recent work is also notable for its geographical coverage, with the results of water quality monitoring programmes being reported from Scandinavia, The Netherlands, Germany, Switzerland and Australia as well as the United Kingdom and the United States. These studies have been representative of a wide range of catchment sizes, types of drainage system, climate and land use. In these circumstances, contradictory results are perhaps to be anticipated. For example, Wilkinson (1956) showed that the strength of pollutants tended to increase with the length of the antecedent dry period. This conclusion was supported by Hedley and King (1971) with regard to the total loadings of both BOD and SS from an older type of development of comparable size in the English Midlands. However, Weibel *et al.* (1964), Bryan (1972), Whipple *et al.* (1977) and Bedient *et al.* (1980), all working in different parts of the United States, have been unable to identify any such relationship. The latter authors attributed the lack of correlation between loadings and duration of the antecedent dry period to differences in climate, soil type and slopes.

Conflicting results such as these compound the already considerable problems of interpreting the variations of urban water quality in both time and space. A better understanding of the physical processes which govern the movement of pollutants within a drainage system and its tributary catchment would greatly assist in resolving some of these difficulties, but as Steele *et al.* (1980) have remarked, research on this topic has been woefully inadequate. Nevertheless, the basis for possible future developments may be found in studies such as those of Cordery (1977), Das (1977), Ellis (1976, 1977), Randall *et al.* (1977), Bedient *et al.* (1978) and Griffin *et al.* (1980), whose results have been presented in the form of 'chemographs' or 'pollutographs' showing the variation with time of individual water quality constituents during isolated storm events. A particular feature of such diagrams is the rapid increase in concentration which occurs at the beginning of a storm event. According to Griffin *et al.* (1980), the presence of this 'first flush' phenomenon depends upon whether or not the contaminant is in the soluble or the particulate phase. Insoluble pollutants are removed primarily by mechanical processes, whereas soluble components are governed by solubility equilibria and remain largely unaffected.

This approach, which involves the collection of sequential samples throughout the duration of the hydrograph, may be contrasted with that adopted by the majority of previous investigators who have confined themselves to the broader variations of pollutant concentrations and loading rates. Owing to the flow-dependent nature of non-point sources of pollution, loading rates are to be preferred to concentrations for purposes of comparison. Accordingly, in order to illustrate the order of magnitude of variation in the water quality constituents of urban areas, the annual loading rates for BOD, SS, TP and TN obtained from 14 different urban catchments have been summarised in Table 10.1.

The catchment areas from which the data listed in Table 10.1 were obtained are broadly representative of the residential, commercial, and mixed residential and commercial developments to be found in many countries. Nevertheless, the most notable feature of the loading rates quoted in Table 10.1 for all four contaminants is their extreme variability. A proportion of this variance may be accounted for by the procedures used to extrapolate the loadings observed in individual

**Table 10.1. Comparison between reported loading rates (kg/ha/year) for BOD, SS, TP and TN from different urban areas**

| *Source* | *Location (land use)* | *Area (ha)* | *Loading rate (kg/ha/year)* | | | |
|---|---|---|---|---|---|---|
| | | | *BOD* | *SS* | *TP* | *TN* |
| Wilkinson (1956) | Oxhey, UK (resid.) | 247 | 4·6 | 115 | — | — |
| Weibel *et al.* (1964) | Cincinnati, USA (resid./commerc.) | 11 | 37 | 818 | — | 10 |
| Bryan (1972) | Durham, USA (resid./commerc.) | 432 | 94 | — | — | — |
| Soderlund and Lehtinen (1973) | Stockholm, Sweden (road interchange) | 3·3 | 100 | 1185 | 0·2 | 5·1 |
| | (resid.) | 25 | 14 | 173 | 0·04 | 1·4 |
| | (commerc.) | 25 | 43 | 620 | 0·16 | 3·5 |
| Hedley and Lockley (1975) | Birmingham, UK (motorway) | 3·6 | 172 | 6289 | — | — |
| Randall *et al.* (1977) | Massanas, USA (resid.) | 80 | — | 560 | 0·82 | 2·9 |
| | (resid./commerc.) | 1680 | 11·5 | 1787 | 1·37 | 5·5 |
| Roberts *et al.* (1977) | Zurich, Switzerland (resid.) | 9 | 25 | 159 | 0·59 | 4·9 |
| Cordery (1977) | Sydney, Australia (resid.) | 131 | 35 | 320 | — | — |
| Wanielista *et al.* (1977) | Orlando, USA (resid.) | 6·5 | 75 | 195 | — | 1·8 |
| | (commerc.) | 11·4 | 50 | 338 | — | 4·0 |
| Bedient *et al.* (1978) | Houston, USA (resid. under construction) | 8744 | — | 2676 | 0·75 | — |

storms to the annual level. In order to explain the remainder, a more detailed consideration of the characteristics of each type of development is necessary.

The highest loading rates for SS and BOD in Table 10.1 are those for a length of partly elevated motorway quoted by Hedley and Lockley (1975). These figures are thought to be associated with the addition of de-icing salts in winter and the pollution products from over 400 000 motor vehicles per week. In general terms, the loading rates for lengths of road are seen to be higher than those for commercial or mixed commercial and residential areas. The exception is the result quoted by Bedient *et al.* (1978) for an area under construction. Since 74% of the catchment remained under forest cover, which would provide effective erosion control, this figure is particularly significant. In turn, mixed commercial and residential areas appear to yield larger loading rates than those from purely residential areas. This ranking in terms of land use tends to suggest that vehicle numbers, or perhaps the area covered by roads, may have an important influence on pollution levels. Support for this argument may be obtained from the results of separate studies into the quality of stormwater runoff within street gutters prior to entry into the sewer (see, for example, Waller, 1972; Sartor *et al.*, 1974).

Using data from 10 sites representing a variety of urban land uses, Waller (1972) showed that average SS concentrations were generally higher in commercial than in residential areas, but that mean BOD concentrations tended to be higher in residential than in commercial subdivisions. Waller also noted that peak SS concentrations were relatively unaffected by the length of the antecedent dry period, a result that was attributed to frequent street sweeping. However, Sartor *et al.* (1974) found that conventional street-cleaning methods were relatively inefficient in removing the fine solids fraction of street surface contaminants which may account for up to a quarter of the oxygen demand, one-third to one-half of the algal nutrients and over half the heavy metals in less than one-tenth of the total solids loading. Those authors also showed that the loading of pollutants was some 80% larger on asphalt streets than on concrete roads, and that a deterioration in the condition of the carriageway could increase the loading by a factor of up to 250%.

The variability in BOD and SS shown in Table 10.1 is also evident for both TP and TN. As noted in Section 10.2, the levels of these constituents have an important bearing on algal productivity in water bodies such as lakes and reservoirs. According to Sawyer (1947), nuisance blooms of algae can be expected to grow when levels of inorganic N exceed 0·3 mg/litre, organic P exceed 0·1 mg/litre and inorganic P exceed 0·01 mg/litre. The results shown in Table 10.1 are therefore sufficient to indicate that the control of eutrophication is unlikely to be effective if attention is confined to controlling point sources of pollutants, as emphasised by Randall *et al.* (1978).

The figures for TN and TP given in Table 10.1 are broadly in agreement with the loading rates for urban drainage of 0·9 kg/ha/year P and 9·5 kg/ha/year N quoted by Owens (1970) in connection with a study of the Great Ouse river basin in England. The source of these nutrients in urban runoff has been the subject of some controversy. For many years, fertilisers were thought to be the principal source of phosphorus in stormwater from urban areas. However, Cowen and Lee (1973) have shown that, depending upon the species, up to one-fifth of the phosphorus content of leaves may be leached by runoff water, which suggests that vegetal material may be a more significant source of nutrients than fertiliser. Confirmation of this finding has recently been provided by Waller (1977), who also advocated the modelling of individual sources and flows of phosphorus within a drainage area as a basis for estimating the composition of the runoff.

The above discussion serves to illustrate the large number of variables which can influence the loading rates of pollutants found in urban stormwater runoff. Even if the loading rates for unmonitored catchments could be estimated with any reliability, the changes in the quality of the receiving watercourse depend upon its capacity both to assimilate and to transport contaminants. Calibration data are therefore fundamental to the success of any attempt to model the changes in water quality in both the runoff from individual drainage areas and the receiving water body. As underlined by Wanielista *et al.* (1977) and Daniel *et al.* (1978) among others, few planning studies that involve water quality considerations can be regarded as complete without an associated programme of field measurements.

Although the data summarised in Table 10.1 may pose more questions than they answer on the character of non-point source pollution, they are sufficient to establish its importance in relation to water quality management. Information on this aspect has been provided by several authors (including Soderlund and Lehtinen, 1973; Randall *et al.*, 1977; Roberts *et al.*, 1977; Cordery, 1977; Hajas *et al.*, 1978; Pirner and Harms, 1978; and Field and Turkeltaub, 1980), who have followed Wilkinson (1956) and Weibel *et al.* (1964) in comparing the composition of urban stormwater with that of the effluent from a sewage treatment works. The results from these comparisons have shown a gratifying measure of agreement, with the urban runoff producing from 2 to 6 times or more the loading of SS, somewhat less BOD and substantially smaller amounts of TP and TN than the treated effluent. These findings demonstrate further the fallacy of confining water quality control programmes to the improvement of recorded point source discharges.

Finally, although this discussion has been devoted to predominantly urban land uses, the pollution potential of agricultural non-point sources should not be ignored. Yu *et al.* (1975) have pointed out that agricultural areas can contribute as much, if not more, organic pollution than an urban residential area. With regard to nutrient loadings, both Sawyer (1947) and Owens (1970) have quoted annual rates for the phosphorus and nitrogen in agricultural drainage comparable in magnitude to urban runoff. Table 10.2, which was prepared by Wanielista *et al.* (1977) using data from a further 14 references, compares the annual loading rates of BOD, SS, TN and P obtained from four different types of land use. Although the variability of the results is comparable to that of Table 10.1, these data show that cultivated land and, to a lesser extent, pasture may be a significant source of both SS and nutrients. The case for monitoring the water quality of cropped areas is therefore as convincing from a pollution control viewpoint as that for urban catchments.

**Table 10.2. Averages and ranges of loading rates (kg/ha/year) for four different types of land use**
(After Wanielista *et al.*, 1977, by permission)

| *Land use* | *Statistic* | *BOD* | *SS* | *P* | *TN* |
|---|---|---|---|---|---|
| Urban | Average | 75 | 1700 | 2·0 | 8·5 |
| | Range | 53–82 | 728–4794 | 1·0–5·0 | 5·2–18 |
| Pasture | Average | 11 | 840[a] | 0·3 | 5·3 |
| | Range | 6–17 | 11·8–840 | 0·24–0·66 | 2·5–8·5 |
| Cultivated | Average | 18 | 4200[a] | 1·05 | 26 |
| | Range | 4–31 | 286–4200 | 0·18–1·62 | 15–37 |
| Woodland | Average | 5 | 98 | 0·1 | 3·1 |
| | Range | 4–7 | 45–432 | 0·01–0·86 | 2·4–5·1 |

[a] The larger in a sample of 2 values.

## REFERENCES

ANGINO, E. E., MAGNUSON, L. M. and STEWART, G. F. (1972) Effects of urbanisation on storm water quality: a limited experiment, Naismith Ditch, Lawrence, Kansas. *Wat. Resour. Res.*, **8,** 135–40.

BARKDOLL, M. P., OVERTON, D. E. and BETSON, R. P. (1977) Some effects of dustfall on urban stormwater quality. *J. Wat. Poll. Contr. Fed.*, **49,** 1976–84.

BARTON, B. A. (1977) Short-term effects of highway construction on the limnology of a small stream in southern Ontario. *Freshwater Biol.*, **7,** 99–108.

BEDIENT, P. B., HARNED, D. A. and CHARACKLIS, W. G. (1978) Storm-water analysis and prediction in Houston. *Proc. Am. Soc. Civ. Engrs., J. Environ. Engng. Div.*, **104**(EE6), 1087–1100.

BEDIENT, P. B., LAMBERT, J. L. and SPRINGER, N. K. (1980) Stormwater pollutant load–runoff relationships. *J. Wat. Poll. Contr. Fed.*, **52,** 2396–2404.

BETSON, R. P. (1978) Bulk precipitation and streamflow quality relationships in an urban area. *Wat. Resour. Res.*, **14,** 1165–9.

BRYAN, E. H. (1972) Quality of stormwater drainage from urban land. *Wat. Resour. Bull.*, **8,** 578–88.

BRYAN, E. H. (1974) Concentrations of lead in urban storm water. *J. Wat. Poll. Contr. Fed.*, **46,** 2419–21.

BURT, T. P. and DAY, M. R. (1977) Spatial variations in rainfall and stream water quality around the Avonmouth industrial complex. *Int. J. Environ. Studies*, **11,** 205–9.

CHARACKLIS, W. G., WARD, C. H., KING, J. M. and ROE, F. L. (1979) Rainfall quality, land use and run-off quality. *Proc. Am. Soc. Civ. Engrs., J. Environ. Engng. Div.*, **105**(EE2), 416–19.

CORDERY, I. (1977) Quality characteristics of urban storm water in Sydney, Australia. *Wat. Resour. Res.*, **13,** 197–202.

COWEN, W. F. and LEE, G. F. (1973) Leaves as a source of phosphorus. *Environ. Sci. Technol.*, **7,** 853–4.

DANIEL, T. C., MCGUIRE, P. E., BUBENZER, G. D., MADISON, F. W. and KONRAD, J. G. (1978) Assessing the pollutional load from non-point sources: planning considerations and a description of an automated water quality monitoring programme. *Environ. Management*, **2,** 55–65.

DAS, K. C. (1977) Quality of combined sewer overflows from urban and semi-urban areas in Richmond, Virginia, USA. Int. Assoc. Hydrol. Sci., Publ. No. 123, pp. 265–76.

DETHIER, D. P. (1979) Atmospheric contributions to stream water chemistry in the North Cascade Range, Washington. *Wat. Resour. Res.*, **15,** 787–94.

ELLIS, J. B. (1976) Sediments and water quality of urban storm water. *Wat. Serv.*, **80,** 730–4.

ELLIS, J. B. (1977) The characterisation of particulate solids and quality of water discharged from an urban area. Int. Assoc. Hydrol. Sci., Publ. No. 123, pp. 283–91.

FIELD, R. and TURKELTAUB, R. (1980) Don't underestimate urban run-off problems. *Wat. and Wastes Engng.*, **17**(10), 48–52.

GRAF, W. H. (1971) *Hydraulics of sediment transport* (McGraw-Hill, New York) 513 pp.

GRIFFIN, D. M., GIZZARD, T. J., RANDALL, C. W., HELSEL, D. R. and HARTIGAN, J. P. (1980) Analysis of non-point pollution export from small catchments. *J. Wat. Poll. Contr. Fed.*, **52,** 780–90.

HAJAS, L., BOGGS, D. B., SHUCKROW, A. J. and QUIMPO, R. G. (1978) Projecting urban run-off flows and loads. *Proc. Am. Soc. Civ. Engrs., J. Environ. Engng. Div.*, **104**(EE6), 1149–63.

HEDLEY, G. and KING, M. V. (1971) Suggested correlation between storm sewage characteristics and storm overflow performance. *Proc. Instn. Civ. Engrs.*, **48,** 399–411.

HEDLEY, G. and LOCKLEY, J. C. (1975) Quality of water discharged from an urban motorway. *Wat. Poll. Contr.*, **74**(6), 659–74.

HUFF, F. A. (1976) Relation between atmospheric pollution, precipitation and stream-water quality near a large urban–industrial complex. *Wat. Res.*, **10,** 945–53.

HUNTER, J. V., SABATINO, T., GOMPERTS, R. and MACKENZIE, M. J. (1979) Contribution of urban runoff to hydrocarbon pollution. *J. Wat. Poll. Contr. Fed.*, **51,** 2129–38.

JAMES, A. (1976) Water quality. In Rodda, J. C. (ed.), *Facets of hydrology* (Wiley, New York) pp. 177–98.

JANARDAN, K. G. and SCHAEFFER, D. J. (1981) Methods for estimating the number of identifiable organic pollutants in the aquatic environment. *Wat. Resour. Res.*, **17,** 243–9.

KARR, J. R. and SCHLOSSER, I. J. (1978) Water resources and the land–water interface. *Science*, **201,** 229–34.

LAXEN, D. P. H. and HARRISON, R. M. (1977) The highway as a source of water pollution: an appraisal with the heavy metal lead. *Wat. Res.*, **11,** 1–11.

LEWIS, W. M. (1981) Precipitation chemistry and nutrient loading by precipitation in a tropical watershed. *Wat. Resour. Res.*, **17,** 169–81.

LEWIS, W. M. and GRANT, M. C. (1978) Sampling and chemical interpretation of precipitation for mass balance studies. *Wat. Resour. Res.*, **14,** 1098–1104.

MARSH, T. J. (1980) Towards a nitrate balance for England and Wales. *Wat. Serv.*, **84,** 601–6.

MURPHY, C. B. and CARLEO, D. J. (1978) The contribution of mercury and chlorinated organics from urban runoff. *Wat. Res.*, **12,** 531–3.

OWENS, M. (1970) Nutrient balances in rivers. *Wat. Treatment and Exam.*, **19,** 239–52.

PIRNER, S. M. and HARMS, L. L. (1978) Rapid City combats the effects of urban run-off on surface water. *Wat. and Sewage Wks.*, **125**(2), 48–53.

QURESHI, A. A. and DUTKA, B. J. (1979) Microbiological studies on the quality of urban storm water run-off in southern Ontario, Canada. *Wat. Res.*, **13,** 977–85.

RANDALL, C. W., GARLAND, J. A., GIZZARD, T. J. and HOEHN, R. C. (1977) The significance of storm-water run-off in an urbanised watershed. *Progr. in Wat. Technol.*, **9,** 547–62.

RANDALL, C. W., GIZZARD, T. J. and HOEHN, R. C. (1978) Effect of upstream control on a water supply reservoir. *J. Wat. Poll. Contr. Fed.*, **50,** 2687–2702.

ROBERTS, P. V., DAUBER, L., NOVAK, B. and ZOBRIST, J. (1977) Pollutant loadings in urban storm water. *Progr. in Wat. Technol.*, **8**(6), 93–101.

SARTOR, J. D., BOYD, G. B. and AGARDY, F. J. (1974) Water pollution aspects of street surface contaminants. *J. Wat. Poll. Contr. Fed.*, **46,** 458–67.

SAWYER, C. N. (1947) Fertilisation of lakes by agriculture and urban drainage. *J. New Engl. Wat. Wks. Assoc.*, **61,** 109–27.

SCOTTISH DEVELOPMENT DEPARTMENT (1977) *Storm sewage: separation and disposal.* Rept. of the Working Party on Storm Sewage (Scotland) (HMSO, London) 120 pp.

SIDWICK, J. M. (1977) *A brief history of sewage treatment* (Thunderbird Enterprises Ltd, Harrow, UK) 68 pp.

SODERLUND, G. and LEHTINEN, H. (1973) Comparison of discharges from urban storm-water run-off, mixed storm overflow and treated sewage. In Jenkins, S. H. (ed.), *Advances in Wat. Poll. Res., Proc. 6th Int. Conf.* (Pergamon Press, Oxford) pp. 309–25.

SOLOMON, R. L., HARTFORD, J. W. and MEINKOTH, D. B. (1977) Sources of automotive lead contamination of surface water. *J. Wat. Poll. Contr. Fed.*, **49,** 2502–4.

STEELE, T. D., AVERETT, R. C., DAS, K. C., JONES, B. F., STEFAN, H. G., WARD, R. C. and YU, S. L. (1980) Water quality research: an overview of areas of concern. *Eos, Trans. Am. Geophys. Un.*, **61,** 433–7.

VANONI, V. A. (ed.) (1975) *Sedimentation engineering.* Am. Soc. Civ. Engrs., Manuals and Repts. on Engng. Practice No. 54, 745 pp.

WALLER, D. H. (1972) Factors that influence variations in the composition of urban surface runoff. *Wat. Poll. Res. in Canada*, **7,** 68–95.

WALLER, D. H. (1977) Effects of urbanisation on phosphorus flows in a residential system. Int. Assoc. Hydrol. Sci., Publ. No. 123, pp. 52–8.

WANIELISTA, M. P., YOUSEF, Y. A. and MCLELLAN, W. M. (1977) Non-point source effects on water quality. *J. Wat. Poll. Contr. Fed.*, **49,** 441–51.

WEIBEL, S. R., ANDERSON, R. J. and WOODWARD, R. L. (1964) Urban land runoff as a factor in stream pollution. *J. Wat. Poll. Contr. Fed.*, **36,** 914–29.

WHIPPLE, W. and MCINTOSH, A. W. (1979) Stream surveillance and monitoring in urban areas. *Wat. and Sewage Wks*, **126**(4), 72–5.

WHIPPLE, W., HUNTER, J. V. and YU, S. L. (1974) Unrecorded pollution from urban run-off. *J. Wat. Poll. Contr. Fed.*, **46,** 873–85.

WHIPPLE, W., HUNTER, J. V. and YU, S. L. (1977) Effects of storm frequency on pollution from urban run-off, *J. Wat. Poll. Contr. Fed.*, **49,** 2243–8.

WHITEHEAD, H. G. and FETH, J. H. (1964) Chemical composition of rain, dry fallout, and bulk precipitation at Menlo Park, California, 1957–1959. *J. Geophys. Res.*, **69,** 3319–33.

WIGGERS, J. B. M. and BAKKER, K. (1978) In what way is water pollution influenced by sewerage systems? *Hydrol. Sci. Bull.*, **23,** 257–66.

WILKINSON, R. (1956) The quality of urban rainfall run-off water from a housing estate. *J. Instn. Pub. Health Engrs.*, **55,** 70–84.

YU, S. L, WHIPPLE, W. and HUNTER, J. V. (1975) Assessing unrecorded organic pollution from agricultural, urban and wooded lands. *Wat. Res.*, **9,** 849–52.

# 11

# Water Quality Modelling for Urban Areas

## 11.1. INTRODUCTION

Loucks (1981) has defined a water quality model as a set of mathematical expressions describing the physical, chemical and biological processes that are assumed to take place in a water body. For a given group of point and non-point sources and a specified hydrological regime, such a model provides a set of outputs which consist of values of water quality parameters for each time period at predetermined cross-sections of the watercourse. Since each river system tends to have its own peculiar hydrological characteristics and to pose its own particular pollution problems, the variety of available water quality models is wide. Moreover, owing to the difficulties encountered in describing comprehensively the physical, chemical and biological mechanisms that are involved in the self-purification processes in a water body, many of these models lack generality. This situation has arisen primarily because of what Cembrowicz *et al.* (1978) have referred to as the prohibitive bottlenecks imposed by data availability and the determination of characteristic parameters and constants. Conversely, the detail appropriate to assessing the impact of a point source discharge is not required for long-range planning purposes, so that the need for a range of problem-oriented water quality models is inevitable.

When employed in planning applications, additional mathematical statements are added to the water quality model in order to define the alternative management strategies to be evaluated and the management objectives that are to be minimised or maximised. These models are beyond the scope of the present review, but comprehensive descriptions are available in various symposia, such as that on the Trent

River Basin model (Institution of Water Engineers, 1972) and collections of case studies, such as that edited by Biswas (1981). Attention is concentrated on the more restricted problem of modelling the water quality of urban catchment areas, which may be considered to consist of two distinct parts:

(1) the estimation of storm runoff loadings; and
(2) the description of water quality variations in receiving water bodies.

Despite the variations in both the scope and form of models of stormwater pollutant loadings, the majority adhere to the same set of basic assumptions, among the more frequently encountered of which are that:

(a) the accumulation of solids on the surface of an urban catchment is a function of the elapsed time since the previous storm event or the last street-sweeping operation; and
(b) the loading of a particular pollutant may be estimated from the total solids loading by the application of a ratio known as a 'potency factor' or 'availability factor'.

As discussed in Section 10.3, assumption (a) above remains open to debate, but continues to be applied in the absence of better-substantiated alternatives. The use of potency factors is similarly controversial, since their application implies that the pollutants are conservative. Nevertheless, many authors, including Litwin and Donigian (1978) and Sutherland and McCuen (1978), have argued that this approach is reasonable for insoluble or partially soluble contaminants such as heavy metals, many nutrient forms and organic matter. The popularity of the potency factor concept provides some measure of the impasse that appears to have been reached in the development of water quality models. As noted by Field and Turkeltaub (1980), conventional water quality models exclude sediment transport, and existing sediment transport models consider only discrete sediments and not biochemical reactions. However, some progress has been achieved in describing the interactions between pollutants and particulate matter for particular constituents, as shown by the nutrient model proposed by Novotny *et al.* (1978).

Bearing in mind the restrictive nature of the above-mentioned assumptions, the available approaches for estimating the pollutant loadings carried by stormwater runoff are reviewed in Section 11.2.

The details of two well-known deterministic runoff quality models are then described in Section 11.3. The outputs from such models in the form of storm hydrographs and chemographs may be employed as the inputs to representations of the receiving water system. Here again, a wide variety of both steady-state and transient models is available, whose dimensionality is largely dependent upon the assumptions made about the mixing of pollutants within the water body. For example, two-dimensional models may assume lateral mixing, as in a stratified estuary, or vertical mixing as in a wide-shallow river, and one-dimensional models imply complete mixing in both the vertical and lateral directions. Steady-state models are inevitably simpler and require less computational effort than transient models, but the latter are more relevant to short-term management and control. The basis of such models is outlined briefly in Section 11.4.

## 11.2. ESTIMATION OF STORMWATER POLLUTANT LOADINGS

In discussing the estimation of contaminant loadings resulting from stormwater runoff, Wu and Ahlert (1978) recognised three distinct levels of detail at which information might be required:

(1) the average annual storm load, which is assumed to occur continuously throughout both wet and dry periods but nonetheless provides a convenient measure of the relative impact of surface water runoff compared with point source discharges;
(2) inter-event loading, which takes into account the variations that occur from storm to storm; and
(3) intra-event loading, which considers the transient water quality state during an individual storm event.

Those authors also identified four different types of model that might be applied to satisfy one or more of the above-mentioned requirements:

(i) the zero-order method, also referred to by Litwin and Donigian (1978) as the empirical methods;
(ii) direct methods;
(iii) statistical methods; and
(iv) descriptive methods.

Each of the above is described briefly in the following subsections.

### 11.2.1. Zero-Order (or Empirical) Method

This, the simplest of the above-mentioned approaches, involves the application of unit loading rates for various water quality constituents to the areas with different types of land use within a drainage basin. These unit loading rates should strictly be estimated from local water quality monitoring programmes, but in the absence of such data may be drawn from published work on catchments with similar hydrological characteristics and land use patterns. The potential difficulties of transferring information between drainage areas may be appreciated from the wide variation in annual loading rates for areas with ostensibly the same land use shown in Tables 10.1 and 10.2. Nevertheless, where local data are available, this method of approach should be capable of yielding results that are sufficiently reliable at (say) the pre-feasibility study level.

### 11.2.2. Direct Method

The direct method is based upon the assumption that the average discharge and the mean concentration of a specified pollutant obtained from a given catchment area are independent variables, so that the average loading may be estimated from their product. The mean concentration may again be taken from published sources. Alternatively, its value may be determined from a grab sampling programme as a flow-weighted concentration or estimated by the application of a regression equation derived using method (iii).

### 11.2.3. Statistical Methods

This approach is based upon the application of classical statistical procedures, such as multiple linear regression analysis (MLRA) or discriminant analysis, to water quality data. Where MLRA has been employed, the dependent variables have consisted of total storm event loads or the mean concentrations of various pollutants, and the independent variables have included descriptors of climate, land use, topography and season. A summary of the variables and parameters that may be significant in such an exercise, quoted by Jewell and Adrian (1982), is reproduced in Table 11.1. The 'storm event total dynamic variables' may explain differences between storms for a given catchment, and the 'instantaneous flux dynamic variables' may be relevant to differences within a particular storm.

An attempt to apply the statistical approach to the urban stormwater pollution data published in the United States up to 1972 reported by

**Table 11.1. Independent variables and parameters influencing the washoff of pollutants by stormwater runoff**
(after Jewell and Adrian, 1982, p. 490, by permission)

| *Dynamic variables* | | *Parameters* |
|---|---|---|
| *Storm event totals* | *Instantaneous flux* | |
| Time since last storm event | Runoff intensity | Land use |
| Street-cleaning practices | Cumulative runoff volume | Area |
| Total runoff volume | Time from start of storm | Percent impervious area |
| Storm duration | Rainfall intensity | Overland flow length |
| Total rainfall volume | Cumulative rainfall volume | Percent street and parking area |
| Average rainfall rate | | Street length/unit area |
| | | Population density |
| | | Particulate fallout rate |
| | | Number of catchbasins/unit area |
| | | Climatological data |

Bradford (1977) was largely unsuccessful, a result which the author attributed to unidentified independent variables that explain a high proportion of the variance of the data set. More recently, a similar exercise was undertaken by Jewell and Adrian (1982), whose data base included some 257 storms from 26 drainage basins in 12 geographical regions of the United States. MLRA was used to evaluate 14 models for storm event total loadings and 15 for instantaneous flux. The results obtained showed that no single model provided consistently high explained variances and no one model fitted the majority of catchment areas. Those authors concluded that even assuming the same model to predict the washoff of different pollutants from the same drainage basin may not be warranted and that, wherever possible, local data should be gathered and a representative model derived for each catchment using statistical techniques.

In contrast, more encouraging results were obtained with a statistical approach by Darby *et al.* (1976), who were able to relate a number of catchment 'activity indicators' to observed water quality conditions for drainage areas in Pennsylvania using discriminant analysis. These activity indicators, which included the daily loadings from sewage treatment works, areas of unsewered residential property, population density, developed area and other characteristics obtainable from local government records, could then be used both to estimate the overall water quality and to predict the more troublesome contaminants for ungauged catchments. Those authors also demonstrated the application of this type of information to the design of water quality monitoring programmes.

### 11.2.4. Descriptive Methods

The descriptive methods may be broadly divided into two categories:

(a) loading functions based upon the Universal Soil Loss Equation (USLE) that predict sediment loads from which the loadings of other water quality constituents are obtained by the application of potency factors; and
(b) simulation methods, based upon models of dust and dirt accumulation and their removal by both rainfall and street sweeping, which allow for transient effects within storms.

As succinctly described by Hudson (1981), the essence of the USLE is the reduction of the principal variables which affect the amount of soil erosion to numerical values such that, when all the numbers are multiplied together, the product is the annual soil loss per unit area. The structure of the USLE is illustrated diagrammatically in Fig. 11.1.

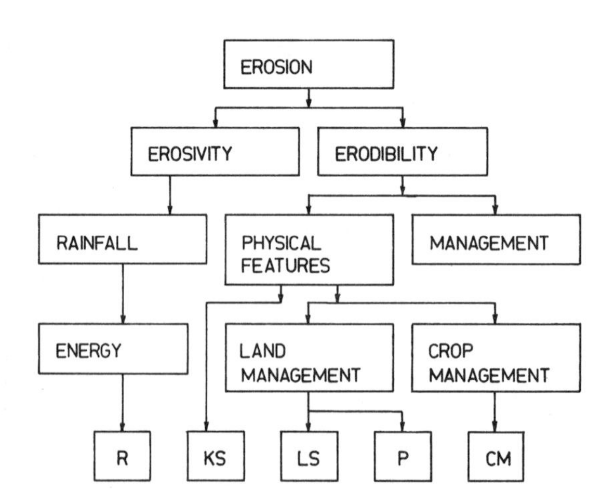


Fig. 11.1. Structure of the Universal Soil Loss Equation (modified from Hudson, 1981, by permission of Batsford Academic and Educational Ltd).

More formally, the USLE may be written as

$$E = R \,.\, KS \,.\, LS \,.\, CM \,.\, P \tag{11.1}$$

where $E$ is the sediment load per unit area; $R$ is the rainfall erosivity index, which is given by the product of the kinetic energy of the storm and the maximum 30 min rainfall intensity within its duration; $KS$ is the soil erodibility factor; $LS$ is a length and slope factor which gives the ratio of soil loss to that of a field with a length of 22·6 m and a slope of 0·09 (the dimensions of the experimental plots from which the factors were enumerated); $CM$ is a crop management factor, which compares the soil loss to that of a cultivated bare fallow field; and $P$ is a conservation practice factor, which compares the soil loss to that of a field with ploughing up and down the steepest slope. Equation (11.1) considers only the soil lost from arable land by splash, sheet and rill erosion; no account is taken of gulley erosion or the degradation of the banks of streams. Estimates of sediment loss obtained by applying the USLE to small catchment areas are therefore subject to some doubt.

Alternatively, the simulation methods depend primarily upon the calibration of expressions for total solids accumulation and removal. Typical forms of model include:

$$S = K_1[1 - \exp(K_2 t)] \tag{11.2}$$

$$S = K_3 t/(1 + K_4 t) \tag{11.3}$$

where $S$ is the loading of total solids/unit length of kerb, $t$ is the time since the last rainfall or the last street-sweeping operation, and $K_i$, $i = 1, 2, 3, 4$, are calibration constants. Another form of model, preferred by Wu and Ahlert (1978) because the abrupt peaks generated by eqns (11.2) and (11.3) may not be representative of loadings from less developed areas, may be written as

$$\ln S_j = a_0 + a_1 S_{j-1} + a_2 T_j + a_3 Q_j \tag{11.4}$$

where $S_j$ is the load at time step $j$, $T_j$ is the time from the beginning of the storm, and $Q_j$ is the discharge at time step $j$.

Expressions of the general form of eqns (11.2) and (11.3) were employed extensively by Sutherland and McCuen (1978) in their Management of Urban Nonpoint Pollution (MUNP) model. A series of equations for accumulation of solids was derived according to land use (single and multiple family residential, commercial and industrial), type and condition of pavement, traffic volume and population. Further expressions were developed for the percentage of solids removed in each of six particle size ranges by both runoff and street-sweeping operations.

## 11.3. URBAN WATER QUALITY MODELS

The different approaches for the estimation of stormwater runoff loadings described above provide only one of the components required for an urban water quality model. Since the simulation of the time series of catchment discharges is a prerequisite to the estimation of pollutant removal by stormwater runoff, such models may be regarded essentially as extended forms of deterministic hydrological model. In some cases, the model may extend to consideration of the receiving water quality, or the treatment of the surface water runoff, or both.

Among the more widely known of the water quality models having such a specification are the Storm Water Management Model (SWMM) (Lager *et al.*, 1971) and the Storage, Treatment and Overflow Runoff Model, STORM (US Army Corps of Engineers, 1974). Of these two models, STORM has by far the simpler hydrological component, but provides continuous simulations of events over an extended time period in contrast to the isolated storms treated in current versions of SWMM. The water quality components of both models are virtually identical.

The overall structure of SWMM has been outlined by Torno (1975), and has already been reviewed briefly in Section 9.4. A comprehensive description of the manner in which pollutant loadings are estimated within recent versions of the model has been provided by Proctor & Redfern Ltd and James F. MacLaren Ltd (1976a,b). Those authors also provided further details of STORM, and so the following notes on both models have been based largely on their work.

Within SWMM, the initial mass of dust and dirt accumulating on the surface of each subcatchment is estimated as a function of the number of preceding dry days and both the frequency and efficiency of street sweeping. The amount of individual contaminants present in the dust and dirt is then established from the land use of the subcatchment. The amount of each pollutant that is washed off the surface during each time interval is assumed to be proportional to the amount which remains, giving an exponential decay function. The rate of removal is controlled by the stormwater discharge and the exponent of the exponential function. Recent versions of SWMM give two optional methods for the estimation of SS loading. The first approach employs an availability factor, which is a function of the rate of runoff in order to account for the lower removal rate of solids at discharges less than the equivalent of 18 mm/h. The second method involves the use of an empirical equation in which the SS removed are expressed as a function of the ratio of the amount remaining at the beginning of the time step to the initial amount at the beginning of the storm. This equation also incorporates a 'removal coefficient' whose value reduces with time over a period of about 5 h. An allowance for the BOD loading associated with SS is made by increasing the BOD by 5% of the weight of SS.

After the contaminant loadings removed from the surface have been computed, attention is turned to the catchbasins or gulley pots. The proportion of solids passing through the latter is determined on the basis of settling velocities such that only the fraction with velocities less than a specified critical figure will be retained. The removal of soluble pollutants is estimated by means of an exponential function of the ratio of the accumulated inflow to the volume of liquid trapped in the gulley pot. Allowances are also made for the subsequent deposition, scour or decay of pollutants in the sewer system. Again, the accumulation of solids is computed as a function of the number of days since the previous storm.

Since SWMM treats only isolated storm events, its use may become prohibitively expensive when attempting to compile statistical information on the accumulation and depletion of pollutants over an extended period. However, STORM, with its simpler hydrological component, is capable of meeting this requirement both conveniently and economically.

STORM operates in hourly time steps using the rainfall recorded at a single raingauge, reduced to a catchment average depth by the application of an areal reduction factor where appropriate, as the input. Rainfall depths exceeding the contents of a depression storage are converted into direct runoff by means of a runoff coefficient. Between rainfall events, depression storage is depleted at the potential evaporation rate. Pervious and impervious areas are considered separately, and allowance is made for snowmelt, but no routing of direct runoff is carried out. The water quality routines are virtually identical to those of SWMM. STORM also allows for the effects of various combinations of storage and treatment of the runoff to be modelled, and their consequences in terms of stormwater overflows to be investigated. No storage is allowed until the specified treatment rate is exceeded. When the storage is completely filled, the excess overflows to the receiving water course. Conversely, when the runoff rate falls below the treatment rate, the spare capacity is used to deplete the storage. The loadings of pollutants in the overflow are determined from a mass balance involving the amounts in the runoff and the storage, and the loss during treatment.

The computer programs for both STORM and SWMM are non-proprietary and readily accessible to potential users. A User's Group has been established for SWMM, which acts as a clearing house for the exchange of information and experience with the model. Since the two models are largely complementary, Proctor & Redfern Ltd and James F. MacLaren Ltd (1976a) have suggested that STORM might be used as a screening model for design events which could then be analysed in detail using SWMM.

## 11.4. RECEIVING WATER QUALITY MODELS

In general, the time scale of receiving water response is very much longer than that of urbanised catchment areas, so that, in modelling their water quality variations, steady-state assumptions are frequently applied. Moreover, the need for numerically tractable formulations with feasible data requirements usually results in the differential equation describing the change in concentration, $C$, of a pollutant with time, $t$, and distance, $X$, downstream of its source being written in the one-dimensional form

$$\frac{\partial C}{\partial t}-\frac{1}{A}\frac{\partial}{\partial X}\left(D_L A\frac{\partial C}{\partial X}\right)+\frac{1}{A}\frac{\partial}{\partial X}(CQ)\pm\sum_k S_k=0 \tag{11.5}$$

where $Q$ is the net downstream freshwater flow, $A$ is the cross-sectional area at the point $X$, $D_L$ is the dispersion coefficient, and $S_k$ is a source (or sink) of the pollutant.

Equation (11.5) states that the change in concentration with respect to time is given by the changes caused by longitudinal dispersion (second term) and advection (third term) plus any sources and minus any sinks. The relative importance of the individual terms is largely dependent upon whether the pollutant under consideration may be classified as either conservative (non-degradable) or non-conservative (degradable). For most practical purposes, this distinction is contingent upon the length of time that the substance remains in the system relative to its rate of decay. With a conservative pollutant, the concentration is reduced only by dilution, and for a single point-source discharge, the fourth term in eqn (11.5) is set to zero. However, with a non-conservative constituent the reactions affecting the increase or decrease in concentration are frequently represented by a first-order process in which the reaction rates are directly proportional to pollutant concentrations, and the fourth term in eqn (11.5) is given by the product $KC$, where $K$ is a rate coefficient. In either case, the dispersion term is usually included only when tidal influences are being considered.

Owing to its fundamental importance to aquatic animal and plant life, oxygen is invariably considered as the primary indicator in water quality modelling studies. The waters of a natural river system are generally close to being saturated with dissolved oxygen (DO). A pollutant discharged into such waters undergoes aerobic decomposition. If the supply of oxygen is unable to match the BOD of the pollutant loading, the receiving water and the type of decomposition become anaerobic. The interaction between the processes of deoxygenation exerted by the BOD and the reoxygenation resulting from the absorption of oxygen from the atmosphere at the water surface and from green plants during photosynthesis produces a time variation of DO referred to as the dissolved oxygen sag curve. A mathematical description of this sag curve was first proposed in 1925 by Streeter and Phelps in their water pollution study of the Ohio River (see Fair *et al.*, 1968, section 33-12).

Denoting the BOD loading as $L$ and the DO deficit below saturation in the flow as $D$, the differential equations describing the two processes may be written as

$$\mathrm{d}L/\mathrm{d}t = -k_1 L \tag{11.6}$$

$$\mathrm{d}D/\mathrm{d}t = k_1 L - k_2 D \tag{11.7}$$

where $k_1$ and $k_2$ are the rate coefficients for deoxygenation and reaeration respectively. The solutions to these equations are

$$L = L_0 \exp(-k_1 t) \tag{11.8}$$

$$D = D_0 \exp(-k_2 t) + KL_0[\exp(-k_1 t) - \exp(-k_2 t)] \tag{11.9}$$

where $L_0$ and $D_0$ are the BOD load and the DO deficit at the discharge point of the pollutant, and $K = k_1/(k_2 - k_1)$.

Equations (11.8) and (11.9) are referred to as the Streeter–Phelps equations and are strictly applicable only if the DO concentration does not fall to zero, in which case the biochemical reaction would change from being aerobic to anaerobic. The DO sag curve is described by eqn (11.9). Of particular interest are the magnitude of the critical DO deficit, $D_c$, and the time of its occurrence, $t_c$, which may be computed by setting $dD/dt$ to zero:

$$D_c = (k_1 L_0/k_2) \exp(-k_1 t_c) \tag{11.10}$$

$$t_c = (k_2 - k_1)^{-1} \ln[(k_2/k_1)\{1 - (D_0/L_0 K)\}] \tag{11.11}$$

The basic Streeter–Phelps equations have been extended by allowance for additional processes, such as the removal or addition of BOD by sedimentation or scouring from the bed of the watercourse, and oxygen production by photosynthesis and oxygen consumption by respiration, a summary of which was presented by Dobbins (1964). In addition, since BOD is composed of several components, each of which displays a different decay pattern, the BOD load may be divided into separate parts, the carbonaceous and the nitrogenous components being the most frequently considered.

As noted by Pentland *et al.* (1972) and Loucks (1981) among others, the Streeter–Phelps equations have provided the basis for the majority of receiving water quality models currently in use. Where the number of reaches is small, DO predictions may be readily obtained using either nomographs (McBride, 1982) or a microcomputer (see Hughto and Schreiber, 1982; Tebbutt, 1982). However, for larger networks and more elaborate problems, recourse must be made to a larger machine and more comprehensive programs such as QUAL II (Grenney *et al.*, 1978; Loucks, 1981) or URBODO (Moodie, 1979). The RECEIV model incorporated into SWMM is applicable only when the major transport mechanism is advection (see Proctor & Redfern Ltd and James F. MacLaren Ltd, 1976a). A similar restriction applies to the program described by Knowles and Wakeford (1978), which was applied to simulate water quality conditions in the River Blackwater, a tributary of the River Thames, by Casapieri *et al.* (1978). An unsteady water quality model for river networks has been described by Tucci and Chen (1981).

Despite the inherent sampling and measurement errors associated with water quality variables and the complex interactions that govern their variability in both time and space, heavy reliance continues to be placed upon the deterministic modelling approach based upon eqn (11.5). In effect, such models provide forecasts of water quality in terms of mean values and ignore the inherent uncertainty that would be described by the probability distribution about the mean. Determination of the latter requires a statistically based methodology, which incorporates allowances for the unpredictable perturbations in time series of water quality variables. One of the simpler models of this type may be obtained through the application of MLRA with pollutant concentrations as the dependent and flows or transformations of flows as the independent variables. For example, Farrimond and Nelson (1980) have described such a model based upon pentad (5-day total) flows. A more sophisticated stochastic modelling approach, in which a technique known as the Instrumental Variable Method is employed to estimate the parameters of differential equation models of water quality, has been described by Whitehead (1980). The application of this procedure to model BOD and DO in a single reach of non-tidal river has been outlined by Whitehead (1977) and Whitehead and Young (1979). With the aid of Monte Carlo (random sampling) experiments using a water quality model of the Bedford Ouse River, Whitehead and Young (1979) were able to show that *a priori* assumed parameter values can be highly misleading. Those authors caution against unreserved acceptance of the outputs from deterministic simulation models, and recommend a thorough sensitivity analysis of the results that they produce.

## REFERENCES

BISWAS, A. K. (ed.) (1981) *Models for water quality management* (McGraw-Hill, London) 348 pp.

BRADFORD, W. L. (1977) Urban stormwater pollutant loadings: a statistical summary through 1972. *J. Wat. Poll. Contr. Fed.*, **49,** 613–22.

CASAPIERI, P., FOX, T. M., OWERS, P. and THOMSON, G. D. (1978) A mathematical deterministic river-quality model. Part 2: Use in evaluating the water-quality management of the Blackwater catchment. *Wat. Res.*, **12,** 1155–61.

CEMBROWICZ, R. G., HAHN, H. H., PLATE, E. J. and SCHULTZ, G. A. (1978) Aspects of present hydrological and water-quality modelling. *Ecol. Mod.*, **5,** 39–66.

DARBY, W. P., MCMICHAEL, F. C. and DUNLAP, R. W. (1976) Urban watershed management using activity indicators to predict water quality. *Wat. Resour. Res.*, **12,** 245–52.

DOBBINS, W. E. (1964) BOD and oxygen relationship in streams. *Proc. Am. Soc. Civ. Engrs., J. San. Engng. Div.*, **90**(SA3), 53–79.

FAIR, G. M., GEYER, J. C. and OKUN, D. A. (1968) *Water and Wastewater Engineering*: Vol. 1, *Water supply and wastewater removal*; Vol. 2, *Water purification and wastewater treatment and disposal* (Wiley, New York) various pp.

FARRIMOND, M. S. and NELSON, J. A. R. (1980) Flow-driven water quality simulation models. *Wat. Res.*, **14,** 1157–68.

FIELD, R. and TURKELTAUB, R. (1980) Don't underestimate urban run-off problems. *Wat. and Wastes Engng.*, **17**(10), 48–52.

GRENNEY, W. J., TEUSCHER, M. C. and DIXON, L. S. (1978) Characteristics of the solution algorithms for the QUAL II river model. *J. Wat. Poll. Contr. Fed.*, **50,** 151–7.

HUDSON, N. W. (1981). *Soil conservation*, 2nd Edn (Batsford, London) 324 pp.

HUGHTO, R. J. and SCHREIBER, R. P. (1982) Microcomputer water-quality simulation model. *Civ. Engng. (ASCE)*, **53**(3), 58–9.

INSTITUTION OF WATER ENGINEERS (1972) *Symposium on advanced techniques in river basin management: the Trent model research programme*, Proc. Birmingham Symp. (IWE, London) 213 pp.

JEWELL, T. K. and ADRIAN, D. D. (1982) Statistical analysis to derive improved stormwater quality models. *J. Wat. Poll. Contr. Fed.*, **54,** 489–99.

KNOWLES, G. and WAKEFORD, A. C. (1978) A mathematical deterministic river-quality model. Part 1: Formulation and description. *Wat. Res.*, **12,** 1149–53.

LAGER, J. A., SHUBINSKI, R. P. and RUSSELL, L. W. (1971) Development of a simulation model for stormwater management. *J. Wat. Poll. Contr. Fed.*, **43,** 2424–35.

LITWIN, Y. J. and DONIGIAN, A. S. (1978) Continuous simulation of nonpoint pollution. *J. Wat. Poll. Contr. Fed.*, **50,** 2348–61.

LOUCKS, D. P. (1981) Water quality models for river systems. Ch. 1 of Biswas, A. K. (ed.), *Models for water quality management* (McGraw-Hill, London) pp. 1–33.

MCBRIDE, G. B. (1982) Nomographs for rapid solutions for the Streeter–Phelps equations. *J. Wat. Poll. Contr. Fed.*, **54,** 378–84.

MOODIE, A. R. (1979) Modelling of water quality and hydrology in an urban watercourse. Austr. Wat. Resour. Council, Tech. Paper No. 45, 146 pp.

NOVOTNY, V., TRAN, H., SIMSIMAN, G. V. and CHESTERS, G. (1978) Mathematical modelling of land run-off contaminated by phosphorus. *J. Wat. Poll. Contr. Fed.*, **50,** 101–12.

PENTLAND, R. L., REYNOLDS, P. J. and BISWAS, A. K. (1972) Water quality modelling: state-of-the-art. In Biswas, A. K. (ed.), *Proc. Int. Symp. on Modelling Techniques in Water Resource Systems* (Environment Canada, Ottawa) Vol. 3, pp. 481–96.

PROCTOR & REDFERN LTD and J. F. MACLAREN LTD (1976a) *Storm water management model study*, Vol. 1, Environment Canada, Res. Progr. for the Abatement of Munic. Poll., Res. Rept. No. 47, 321 pp.

PROCTOR & REDFERN LTD and J. F. MACLAREN LTD (1976b) *Storm water management model study*, Vol. 2, Environment Canada, Res. Progr. for the Abatement of Munic. Poll., Res. Rept. No. 48, 156 pp.
SUTHERLAND, R. C. and MCCUEN, R. H. (1978) Simulation of urban nonpoint source pollution. *Wat. Resour. Bull.*, **14,** 409–28.
TEBBUTT, T. H. Y. (1982) A microcomputer program for dissolved oxygen predictions. *Pub. Health Engr.*, **10**(2), 87–9.
TORNO, H. C. (1975) A model for assessing impact of storm-water runoff and combined sewer overflows and evaluating pollution abatement alternatives. *Wat. Res.*, **9,** 813–15.
TUCCI, C. E. M. and CHEN, Y. H. (1981) Unsteady water quality model for river network. *Proc. Am. Soc. Civ. Engrs., J. Wat. Resour. Plan. and Man. Div.*, **107**(WR2), 477–93.
UNITED STATES ARMY CORPS OF ENGINEERS (1974) Storage, treatment, overflow, runoff model "STORM". USACE Hydrologic Engng. Center, Davis, California, generalised computer program users' manual, various pp.
WHITEHEAD, P. G. (1977) Water quality models for waste water management. Int. Assoc. Hydrol. Sci., Publ. No. 123, pp. 421–30.
WHITEHEAD, P. G. (1980) An instrumental variable method of estimating differential-equation models of dispersion and water quality in non-tidal rivers. *Ecol. Mod.*, **9,** 1–14.
WHITEHEAD, P. G. and YOUNG, P. (1979) Water quality in river systems: Monte-Carlo analysis. *Wat. Resour. Res.*, **15,** 460–8.
WU, J. S. and AHLERT, R. C. (1978) Assessment of methods for computing storm run-off loads. *Wat. Resour. Bull.*, **14,** 429–39.

# 12

# Stormwater Management

## 12.1. INTRODUCTION

Stormwater management is a title used to describe a group of techniques whose common aim is the mitigation of undesirable effects produced by the quantity and quality of urban runoff. Such techniques may be broadly divided into two categories, involving either structural or non-structural measures. A summary of the more widely applied practices which fall under these headings is presented in Table 12.1. In any urban area a selection of these measures may be in concurrent use. Care is therefore necessary to ensure that initiatives intended for the control of the volume of flood water do not have an adverse effect upon water quality. Similarly, steps to improve the quality of both urban runoff and the receiving watercourses should not interfere with the efficient functioning of drainage systems and flood alleviation works. Nevertheless, solutions to the problems of flood control and pollution control posed by urbanisation have tended to be single- rather than dual-purpose.

As noted by Butler (1972), perhaps the most obvious approach to flood alleviation is to improve the capacity of an existing channel by enlarging its cross-section, reducing its roughness or by constructing a relief channel. However, any alterations to the alignment and cross-section of a natural watercourse must take into consideration the processes which brought about its formation. In any catchment area subject to urbanisation, the increased sediment yields which result from construction activity and the changes in flow regime which accompany the urban growth serve to alter channel geometry to the detriment of the physical and biological environment. Channelisation, by removing meanders and increasing the channel cross-section at

**Table 12.1. Stormwater management practices**

| *Type of measure* | *Quantitative* | *Qualitative* |
|---|---|---|
| Structural | Channelisation<br>Balancing ponds<br>Recharge basins<br>Rooftop storage<br>Porous pavements | Effluent treatment at source<br>Balancing ponds<br>Recharge basins |
| Non-structural | Preservation of local landforms<br>Flood plain zoning | Street sweeping<br>Gulley cleaning<br>Anti-litter legislation<br>Control of de-icing |

bank-full flow, may have similar adverse effects. Its consequences are therefore examined in further detail in Section 12.2.

Where the changes in the magnitude of the design flood caused by urbanisation are too large to be accommodated by channelisation, attention is invariably turned to the provision of some form of balancing storage. In an increasing number of cases, an urban development may only be allowed to proceed if the builder ensures that the post-development runoff does not exceed that which would have been expected under pre-development or natural conditions. Indeed, in many metropolitan areas of the United States, such requirements are incorporated into local planning legislation (see McCuen, 1974, 1979; Talhami, 1980; Debo and Ruby, 1982). In effect, a flood storage pond represents an attempt to replace the natural storage capacity that has been lost through urbanisation. Since the former is concentrated at a single site but the latter was distributed throughout the catchment area, a regional perspective is required in the design of such installations in order to ensure that ponds do not worsen rather than lessen downstream flooding problems. A generalised design procedure for flood storage ponds is discussed in Section 12.3.

Unlike channelisation, flood storage ponds may have a beneficial effect upon water quality by the removal of particulate matter by settlement. Unfortunately, the bulk of the pollutant loading is liable to be carried by the large numbers of small and medium-sized storms, whereas the major flood damages result from the larger, more infrequent events. The design of what Whipple (1979) has referred to as dual-purpose detention basins therefore represents a compromise between the delay and attenuation of the larger flood peaks and the provision of long detention times and high trap efficiencies for pollutants, a problem which is also considered in Section 12.3.

In contrast to flood storage ponds, whose principal functions are to reduce the peak rate of inflow and to redistribute the runoff volume over time, recharge basins are intended to contain the whole of the storm hydrograph for subsequent recharge to underlying aquifers. Such basins are therefore confined to regions that have reasonably permeable surficial deposits and a water table that is sufficiently deep to remain below their floor level. In addition to augmenting local groundwater reserves, recharge basins may also effect considerable savings in the cost of outfall sewers. Long Island, New York, provides a ready example of an area where the advantages of recharge basins have been extensively exploited. According to Aronson and Prill (1977), their numbers have grown from 14 in 1950 to about 700 in 1960 and more than 2200 in 1974. The basins consist of open pits between 3·0 and 4·6 m deep and vary in area from 400 to 121 000 $m^2$. Their effect upon the local water balance has been studied extensively (see, for example, Seaburn, 1970; Seaburn and Aronson, 1974).

Of the structural methods for controlling the quantity of stormwater runoff listed in Table 12.1, the use of rooftop storage and porous pavements is largely confined to more localised applications. The ponding of water on flat roofs, possibly with a thin layer of gravel, will obviously increase design loading and may prove to be uneconomical when compared with other measures. The use of porous pavements in car-parking areas encourages local infiltration, but this approach suffers from the same restrictions as the recharge basin. The design of areas of porous pavement has been considered by Jackson and Ragan (1974).

In general, non-structural stormwater management practices involve some element of either prior planning or continual maintenance. For example, Bonham (1974a,b) has argued extensively for existing landforms to be taken into consideration when planning the layout of a new urban development. In an area with a mature drainage network, the streams meander through natural flood plains located between spurs thrown out from the watershed. The hill slopes located between these ridges and the valley floors are often well drained and provide choice sites for development. Encroachment of the flatter flood plains in the valley floors is thereby avoided, and the need to undertake costly flood alleviation works is greatly reduced. Any new development requires open spaces for amenity and recreational purposes, and their location on the flood plains provides the dual benefits of enhancing the quality of the local environment and preserving washlands for use in times of major flood events.

Of the techniques listed in Table 12.1 for the management of urban water quality, that of effluent treatment at source is perhaps the most obvious structural method, but also the most inflexible because of its inability to cope with rapid changes in runoff rates. In contrast, balancing ponds and recharge basins can serve to control flood flows as well as provide an opportunity for the settlement of waterborne solids. The non-structural methods for water quality management are predominantly concerned with preventing the entry of dust and dirt into the drainage network. Unfortunately, mechanical street sweepers are relatively inefficient at removing the fine solids fraction of street dirt, which has been found to account for a significant proportion of the pollution potential (see Sartor *et al.*, 1974). Road gulleys are similarly ineffective in retaining the finer solids for subsequent removal.

One notable contaminant in urban runoff whose presence is a matter of municipal policy and whose distribution is capable of being regulated is the de-icing salt used for snow clearance. According to Field *et al.* (1973), salts are preferred to abrasives such as sand and cinders because of their greater efficiency in melting snow and ice and the avoidance of subsequent cleaning-up operations. The most popular de-icing salts are sodium chloride and calcium chloride, but, as noted by Roth and Wall (1976), their environmental effects are markedly different. Calcium chloride tends to return to its brine state and be discharged through the drainage system, but sodium chloride reverts more readily to its solid form and is therefore potentially more harmful. Field *et al.* (1973) have also drawn attention to the presence of certain additives in de-icing salts that are capable of generating cyanide by photochemical action, thereby posing additional toxicity problems.

Much of the environmental damage that has been caused by de-icing salts has resulted from excessive applications in pursuit of what has been called the 'bare pavement' policy, the basic objective of which is to provide summer driving conditions in winter. Roth and Wall (1976) have pointed out that the application of 170 kg of salt per km of two-lane highway with a 5 mm thick layer of ice at temperatures between −12 and −4°C will produce salt solutions of from 69 000 to

200 000 mg/litre, compared with the frequently quoted safe level for drinking water of 250 mg/litre. Once temperatures rise, the contaminated snow and ice that accumulates on road verges becomes a source of highly toxic meltwater, impairing both soil fertility and the growth of roadside vegetation. Scott (1976) has shown that the highest salt concentrations in urban runoff occur immediately after the start of a thaw but then fall away rapidly with the increase in snowmelt. Even so, Cherkauer (1975) has reported traces of de-icing salts in the runoff from an urban catchment some 7 months after the last application.

Several authors, including Wulkowicz and Saleem (1974) and Hawkins (1976), have shown by mass balance studies that the salt runoff fraction of urban catchments, i.e. the ratio of chlorides removed by runoff to chlorides applied, is invariably less than unity. The portion which is not carried away accumulates locally, either in the soils or in the groundwater. This accumulation may also be reflected in increasing chloride concentrations in the baseflow from the aquifer during the summer season, as demonstrated by Hawkins (1976) for a small catchment in New York State. Any study of the long-term changes in the chloride concentrations in urban runoff must therefore take into account the residual salts associated with de-icing operations in previous years as well as the applications for the current year.

## 12.2. CHANNELISATION

Although the larger and straighter channels that are constructed to contain urban floodwater may be relatively successful in a hydraulic context, the last decade has seen a growing awareness of the adverse impact that such works can have on the physical and biological environment. An appreciation of the detrimental effects of channelisation must begin with the changes in both the flow regime and runoff water quality that are caused by urbanisation. Chapter 10 has shown that one of the most notable changes in water quality that takes place as a catchment is urbanised is a marked increase in the loading of suspended solids. A greater part of the latter is derived at least initially from the accelerated soil erosion which accompanies general construction activity. For example, Keller (1962) estimated that the sediment yield from an urbanising drainage basin in Maryland was at least 6 times that of an equivalent rural area. In addition, Walling and Gregory (1970) noted that suspended sediment concentrations in the runoff from a small catchment in the south-west of England affected by building activity were between 2 and 10, and occasionally up to 100, times those obtained under undisturbed conditions.

In general, the greatest disruption of vegetal cover and the most extensive remodelling of surface features appears to be associated with road construction. Wolman and Schick (1967) have estimated that the building of divided-lane highways in the United States can produce some $2 \times 10^6$ kg of sediment per linear kilometre. Vice *et al.* (1969) showed that a small catchment in Virginia, about 11% of which was affected by the building of a major freeway intersection, yielded about 10 times the amount of sediment normally expected from cultivated land, some 200 times that from grassland and about 2000 times that from forest. Although precautions are frequently taken to limit erosion during construction by the installation of sediment ponds, mulching and temporary grassing, Burton *et al.* (1976) have found that such measures are not always completely successful. A study of erosion control practices by Weber and Reed (1976) has indicated that the siting of ponds is critical to their performance, with on-site sediment traps being more effective than on-stream basins. Those authors have shown that mulching and seeding do not become effective until vegetation is well established. Burton *et al.* (1976) have also pointed out that much can be done to minimise erosion problems by planning land-clearing operations to avoid months of heavy rainfall.

The increase in suspended sediment brought about by urbanisation may also produce fundamental changes in the downstream channel network. Since the pre-development channels are generally too small to carry the extra sediment load, this material is deposited as flood plain alluvium. A study by Graf (1975) of a small urbanising catchment in Colorado recorded a 270% increase in flood plain area. As the extent of impervious area grows with the progress of urban development, sediment production decreases. However, with the increase in the magnitude of peak discharges and the larger volumes of runoff, the channels then begin to erode through the newly accumulated deposits. According to Leopold *et al.* (1964), a river will construct and maintain a channel to carry without overflow a discharge corresponding to a return period between 1·5 and 2 years. Owing to the changing flow regime brought about by urbanisation, the pre-development bank-full discharge will occur more frequently, and channel enlargement therefore takes place. The extent of such enlargement, which reduces both the visual appeal and the recreational value of a stream as well as incurring costs for remedial work, has been investigated by Hammer (1972) and Hollis and Luckett (1976). Using a sample of 78 catchments in Pennsylvania up to 15 $km^2$ in size, Hammer was able to show that the largest channel enlargements were associated with drainage basins having large areas of either streets with storm sewers or impervious surfaces such as car parks. Similar, but less conclusive, results were obtained for 59 catchments in south-east England by Hollis and Luckett.

The drainage engineer's response to the changes in the channel network brought about by urbanisation has often been eminently predictable. Attempts have been made to improve the capacity of the channel by adjustments to its alignment, slope and cross-section, and to reduce bank erosion and bank instability by lining and the placement of riprap. As the major review by Karr and Schlosser (1978) has amply demonstrated, such channelisation leads to the degradation of water quality as well as having adverse impacts on the physical environment and the biota of the watercourse. The latter are summarised in Fig. 12.1, which has been modified from a diagram presented by the same authors.

Figure 12.1 shows that the four most important consequences of channelisation from the environmental viewpoint are vegetation removal, channel deepening, meander removal and the destruction of pools and riffles. Bankside vegetation plays a particularly important role in regulating stream temperature. The increase in temperatures that results from its removal both decreases the capacity of the stream to hold oxygen and causes changes in the structure of aquatic communities as species with low-temperature optima are unable to survive. Karr and Schlosser (1978) also draw attention to the improvements in water quality that have been effected in logging areas by the maintenance of a buffer zone of vegetation alongside the local watercourses. Such zones can be particularly effective in reducing sediment inflows. Since plant nutrients tend to attach themselves to sediment particles, the increase in nutrient loadings that has been found to accompany road construction activities (see Burton *et al.*, 1976) is also mitigated. The rate at which such suspended solids are converted to soluble forms of nutrients also increases markedly with temperature, so that the shade provided by bankside vegetation reduces their utilisation as well as their availability. In addition, the removal of near-stream vegetation, particularly in the headwaters of river systems, results in a loss of

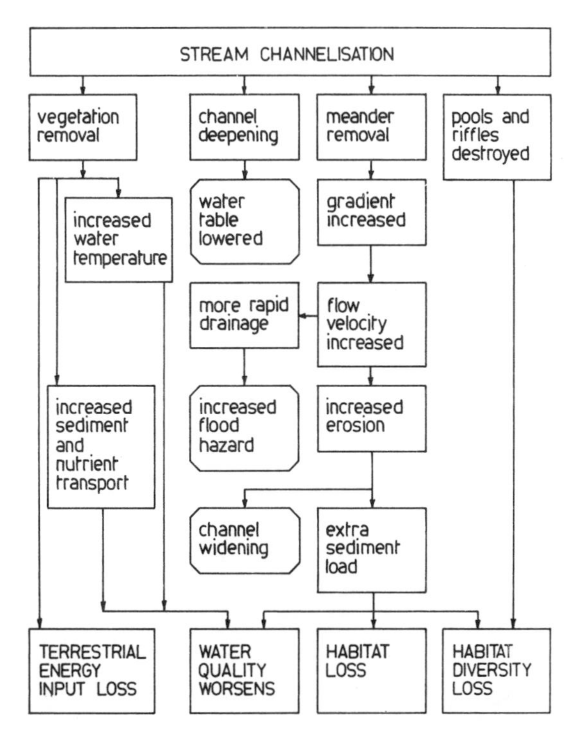


Fig. 12.1. Effects of channelisation in the physical environment and biota of streams (modified from Karr and Schlosser, 1978, by permission of the American Association for the Advancement of Science).

energy inputs that disrupts aquatic food webs and can reduce significantly the production of both invertebrates and fish.

Channel deepening may result from adjustments to bed slope, and can have a localised influence on near-stream groundwater conditions. In contrast, meander removal may have more widespread repercussions. Reducing stream length for the same absolute difference in water surface elevation between two points increases the energy gradient and therefore flow velocities. The more rapid drainage increases flood hazards, but larger sediment particles are also set in motion and erosion is increased, thereby initiating the natural processes by which the channel adjusts its cross-section to the new flow regime. The destruction of riffles and pools in the watercourse may have an effect similar to meander removal upon the energy gradient.

Karr and Schlosser (1978) have also noted a positive correlation between channel sinuosity and the variety of stream habitat, which in turn is directly related to the diversity of the fish population. Even though sediment concentrations may be sub-lethal to fish, the settlement of sediment on spawning grounds and the increased turbidity of the stream clearly alters the structure and productivity of plant, invertebrate and vertebrate communities. For example, Barton (1977) has reported a reduction in the standing crop of fish in an Ontario stream draining a road construction site from 24 to 10 kg/ha.

As an alternative to channelisation, Keller and Hoffman (1976) have advocated a policy of stream restoration, whereby urban channels are altered so that their behaviour is similar to natural watercourses while simultaneously providing some measure of flood alleviation and a visually pleasing appearance. Those authors remark that most urban streams are neglected streams. Stream restoration therefore begins with the removal of urban rubbish, the removal of fallen trees and extensive planting of small trees and brush. In accordance with the principles outlined by Karr and Schlosser (1978), as many trees as possible are left in place. The inside of stream bends is sloped at an angle of 3:1 or less in order to encourage the formation of sandbars. If necessary, the outside of the bend may be protected by riprap in order to ensure bank stability. According to Keller and Hoffman (1976), stream restoration costs less than 5% of an equivalent channelisation scheme, but depends for its ultimate success on the simultaneous enforcement of flood plain zoning and sediment control regulations.

Where channel works cannot be avoided, their design should cause the minimum possible disturbance to the natural drainage system. The design alternatives for developments in mature graded valleys have been examined by Bonham (1974a,b) who recommended the use of flat-vee open channels with rock-lined side slopes set in a relatively flat flood plain. For steep streams, the roughness of the lining would be increased so that the resultant surface instabilities would alert the public to the dangers of the supercritical flows.

## 12.3 FLOOD STORAGE PONDS

Flood storage ponds may be classified according to both their location relative to the river channel (i.e. 'onstream' or 'offstream') and their

**Table 12.2. Classification of flood storage ponds**
(after Hall and Hockin, 1980, by permission of CIRIA)

| *Type of pond* | *Description* |
|---|---|
| Onstream | Dry weather flow passes through the storage area |
| Offstream | Dry weather flow bypasses the storage area |
| Dry | Storage area is free of water under dry weather flow conditions |
| Wet | Storage area contains water under dry weather flow conditions |
| Wet/dry | Part of the storage area contains water and part is free from water under dry weather conditions |

contents under dry weather flow conditions (i.e. 'wet', 'dry' and 'wet/dry'). Definitions of these terms, as presented by Hall and Hockin (1980), are summarised in Table 12.2.

A smaller volume of storage (and therefore less land) is required for an offstream pond than an onstream pond, but more fall is required between the inlet and the outlet if an offstream pond is to be drained by gravity. If the difference in level is insufficient, the depth of the pond may have to be reduced and its area increased, thereby losing the advantage of the saving in land-take.

A flood storage pond may contain water at all times, or may drain either partially or completely between major storm events. A 'wet' pond can provide a useful local amenity for boating or fishing (see, for example, Bunyan, 1975), but such permanent storage obviously cannot be employed to reduce peak discharges. 'Dry' ponds may also be managed as an amenity in the form of playing fields or parkland, but the need to grade the area for drainage purposes may result in some loss of storage capacity.

Flood storage ponds have been used extensively in order to control the surface water runoff from the New Town area of Milton Keynes in central England. According to Davis and Woods (1979), there are to be 12 ponds to serve the area to be developed ultimately, of which 4 will be 'dry'. Particular attention is being paid to the recreational value of the 'wet' and 'wet/dry' ponds, and facilities are to be provided for a variety of pursuits, including yachting, power-boating, canoeing, swimming, fishing, water-skiing and wildlife preservation.

Although the needs of a particular amenity, such as the length of a rowing course or the area available for sailing, could dictate the configuration of a flood storage pond, the principal design criteria are primarily hydrological and hydraulic. In particular, the pond is generally designed to reduce the peak flow rate of a post-development storm hydrograph of a specified return period to a predetermined outflow discharge. The latter usually corresponds to the pre-development peak flow rate from a similar frequency of design storm, but may also be selected on the basis of the capacity of the downstream channel and the available washlands. The extent to which the peak inflow discharge is both delayed and attenuated by an onstream pond is largely determined by the rating curve of the outlet control and the relationship between the volume of storage and the elevation of the water surface of the pond. With an offstream pond, the situation may be further complicated by having a control on the river to limit downstream flow as well as that on the inlet to the pond itself. In either case, the conditions which gave rise to the $T$-year flood prior to construction of the pond will no longer be critical once the works have been commissioned. The peak flow rate is no longer the major consideration because the distribution of the runoff volume within the hydrograph will determine the manner in which the available storage is taken up and therefore the degree of attenuation of the inflow hydrograph. In consequence, the critical storm for the reservoired catchment will have a longer duration, a lower peak but a greater volume of runoff than that for the unreservoired catchment (see also McCuen, 1979; Mein, 1980). The routing of floods corresponding to a range of durations must therefore be carried out in order to identify the critical design storm.

Although flood storage ponds have been used extensively in the New Town areas of the United Kingdom in order to mitigate the downstream effects of the changes in flow regime caused by urbanisation, little advice was available to engineers concerned with their design until the recent publication of a Guide under the auspices of the Construction Industry Research and Information Association (see Hall and Hockin, 1980). In addition to advice on flood estimation methods for catchments subject to urbanisation and techniques for sizing flood storage ponds, this Guide contains a step-by-step design procedure, a brief summary of which is presented in Section 12.3.1. This design procedure is applicable to a single (onstream or offstream) pond. However, flood storage ponds invariably form one element in a larger drainage scheme involving adjacent catchment areas that are also provided with some balancing storage. In such cases, the need arises to evaluate the operation of the system as a whole. The problems of multiple pond systems are therefore examined in Section 12.3.2. Finally, in Section 12.3.3, attention is turned to changes in water quality that can be effected by the action of a flood storage pond.

### 12.3.1. Design Procedure for Flood Storage Ponds

The sequence of steps to be followed in designing a flood storage pond is presented in flowchart form in Fig. 12.2 (see also Hall *et al.*, 1978). For convenience, the procedure may be considered to consist of four parts: the preliminary analysis (steps 1–4); the derivation of flood hydrographs (steps 5–14); the design of the storage pond (steps 15 and 16); and the checking of the pond design (steps 17–20).

The first four steps are intended to enable the designer to form a preliminary assessment of the hydrological significance of urbanisation with a minimum of data collection. In the United Kingdom, the methods recommended in the Flood Studies Report (Natural Environment Research Council, 1975) may be employed to construct a flood frequency distribution for the catchment in its pre-urban state. This distribution may then be modified to reflect the influence of urbanisation using techniques such as those discussed in Section 8.3. The pre-development and post-development discharges for a range of return periods are transformed into stages in order to assess the effect of the changes in terms of frequency of overbank discharges. Where the increase in downstream flows is tolerable or within the capacity of channel improvement works, the use of a flood storage pond need be investigated no further (step 4). However, in order to proceed with the design of a pond, additional data collection is required including a more extensive topographical survey, the results of which can be employed to identify the pond location, flood-prone areas and washlands.

The procedure outlined in Fig. 12.2 is based upon the use of the Unit Hydrograph Method. For ungauged catchments subject to urbanisation, the approach presented by Hall (1981), which has been summarised in Section 8.4, was recommended. This method employs percentage impervious area as the principal measure of the extent of urban development, and particular attention is therefore paid to its assessment in step 6. Once pre-development and post-development

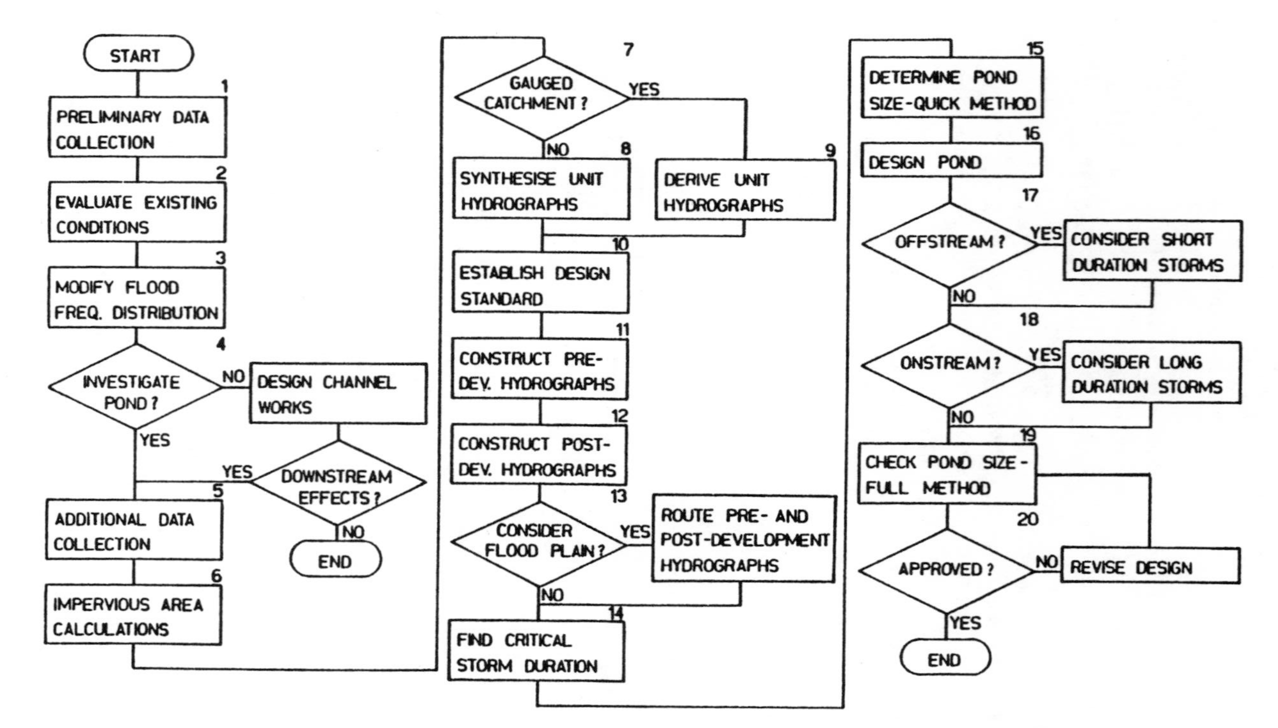


Fig. 12.2 Flowchart for a design procedure for flood storage ponds (modified from Hall *et al.*, 1978, by permission of Pentech Press).

design hydrographs have been synthesised in steps 11 and 12, consideration may be given to the effect of existing washlands, or more probably the loss in overbank storage caused by encroachment of development on to the upstream flood plain, before turning to the determination of the critical storm duration for the pond in step 14.

In many applications the designer may be faced with a wide range of choice in both the size of the pond and the configuration of the control structures. For onstream ponds, the rapid routing techniques described in Section 7.4 may be used to advantage to examine the feasibility of alternative schemes both quickly and economically. For offstream ponds, each possibility must be evaluated individually using finite difference routing methods.

The approach implicit in Fig. 12.2, which allows the simultaneous evaluation of both the storage volume of the pond and the size and layout of the outlet control, has been classified by Donahue *et al.* (1981) as a 'design method'. Those authors drew a distinction between the latter and 'planning methods', which provide only an estimate of the required volume of storage. The majority of the available planning methods are based upon the principles of the Rational Method as outlined in Section 9.2. If the rate of increase in contributing area is assumed to be uniform, and the storm duration, $T_s$, exceeds the time of concentration of the catchment, $T_c$, the inflow hydrograph will assume the characteristic trapezoidal shape shown in Fig. 12.3, with a peak flow rate, $Q$, computed from eqn (9.1). Assuming a constant outflow rate from the pond of $q$, the required storage, $S$, is depicted by the hatched area shown in Fig. 12.3. From the geometry of the inflow and outflow hydrographs,

$$S = QT_s - q(T_s + T_c) + q^2T_c/Q \qquad (12.1)$$

Copas (1957) appears to have been among the first to employ eqn (12.1) to estimate the volume of balancing storage required for overloaded sewerage systems. From eqn (9.1), $Q$ is directly proportional to the average rainfall intensity, which is a function of $T_s$, so that an iterative technique must be used to solve eqn (12.1) for $S$. Similar approaches to that of Copas have been proposed more recently by Kelly (1977), Burton (1980) and Watson (1981). Another technique for estimating storage requirements proposed by Davis (1963) employs the time–area diagram of the catchment to derive the inflow hydrograph. The planning methods proposed by Wycoff and Singh (1976) and Hawley *et al.* (1981) differ from the above in applying regression

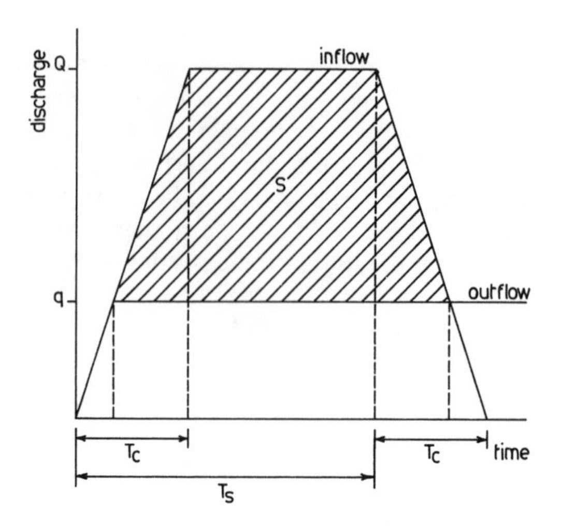


FIG. 12.3. Idealised pond inflow and outflow hydrographs using Rational Method assumptions.

analysis to the results obtained from a full routing method in order to provide preliminary estimates of storage volume or the shape of the outflow hydrograph. Tapp *et al.* (1982) used geometrical approximations to hydrograph shape but incorporated a series of correction factors derived from a full routing technique.

Once the size and configuration of the pond and its control structures have been determined in step 16, a series of check calculations is carried out in which selected storm events are routed through the works as proposed. Since offstream ponds are inlet-controlled, short-duration storms can produce peak flow rates that overload the entrance structure. In contrast, onstream ponds will overdamp the inflows from short-duration storms, but, being outlet-controlled, may produce conditions within long-duration storms in which inflow and outflow are comparable in magnitude and the pond elevation stabilises. This effect may be avoided by either increasing the storage capacity, restricting the outlet control, or both. Should these measures prove to be insufficient, the pond may be changed from onstream to offstream. The design procedure ends in Step 19 with a performance check using a full flood routing method.

### 12.3.2. Multiple Pond Systems

Since the action of a flood storage pond both delays and lowers the peak of the inflow hydrograph, the possibility that an attenuated post-development outflow hydrograph will coincide with and reinforce that from a downstream tributary when the pre-development hydrograph did not, cannot be ignored. A knowledge of such behaviour is particularly important to those schemes involving several ponds in which some form of remote control is employed in order to maximise the use of the available storage capacity as contemplated for Milton Keynes (see Davis and Woods, 1979). The hypothetical example of a cascade of flood storage ponds, in which the outflow from an upstream storage becomes the inflow to its neighbour immediately downstream, has been examined by several authors, including Wycoff and Singh (1976), Mein and Woodhouse (1977) and Mein (1980). Their results show that the reduction in the peak discharge at some downstream point is greater for a single pond than for two ponds, each of half the storage capacity, in series. Similarly, two ponds in series are more effective than three, each providing one-third of the same storage capacity. Apart from its greater hydraulic efficiency, one large pond often has greater recreational potential than several smaller storages. However, a single pond has its maximum effect immediately downstream and its influence diminishes rapidly as the catchment area increases, whereas a cascade of ponds will also reduce the flood peaks over the intermediate lengths of channel.

For the case of ponds in parallel, i.e. located on adjacent tributaries within the same river system, McCuen (1974) and Curtis and McCuen (1977) have described case studies in which conceptual catchment models have been used to show the possibilities for enhancing rather than reducing downstream flood peaks by the construction of storage facilities. However, these increases in peak discharges have tended to be small. Similar studies by Mein and Woodhouse (1977) and Mein (1980) have concluded that the circumstances in which the addition of a flood storage pond enhances peak flows are unusual. Nevertheless, those authors recommend that multiple pond systems should be fully analysed in order to evaluate their overall performance.

To date, few authors have considered the economic aspects of incorporating flood storage ponds into a larger stormwater drainage system, although widespread agreement exists on the importance of allowing for a continuing expenditure on the maintenance of such facilities (McCuen, 1979; Hall and Hockin, 1980; Debo and Ruby, 1982). The application of dynamic programming to find the optimal sizes and locations for flood storage ponds within a catchment area for minimum cost by Mays and Bedient (1982) is therefore a welcome innovation.

### 12.3.3. Water Quality Considerations

As flow enters a flood storage pond, its velocity reduces markedly such that the bulk of any sediment being transported is deposited either within the pond or the region of backwater influence in the approach channel. Unless arrangements are made for its removal, the accumulation of sediment may ultimately encroach on the storage required for the functional operation of the pond. Since many pollutants, such as nutrients and heavy metals, are sediment-related, as noted in Section 10.1, their retention can have a significant influence in maintaining, if not improving, river water quality downstream of an urban area. These secondary benefits of constructing flood storage ponds may be particularly important during the early phases of urban development when the natural vegetal cover is removed and earth-moving is taking place.

The ability of a pond to retain sediment is characterised by its trap efficiency, TE, which is defined as the ratio of the sediment deposited to the total sediment inflow. Trap efficiency is determined primarily by the sediment particle fall velocity, which depends on such factors as particle size and shape, the viscosity and chemical composition of the water, and the rate of flow through the pond, which is a function of the volumes of inflow and available storage and the rate of outflow. Owing to the complexity of these relationships, TE is generally estimated by empirical methods. For example, using data from 44 reservoirs, Brune (1953) established a semi-logarithmic relationship between trap efficiency and the quotient of reservoir capacity, $S$, and average annual inflow, $I$. The parameter $S/I$, which reflects the average number of times that the stored water is replaced during the year, was used as the independent variable in a regression analysis of data from 20 reservoirs with catchment areas less than 39 $km^2$ by Heinemann (1981), which gave the equation

$$\mathrm{TE} = [119{\cdot}6(S/I)/\{0{\cdot}012 + 1{\cdot}02(S/I)\}] - 22{\cdot}0 \qquad (12.2)$$

where TE is expressed as a percentage and $S$ and $I$ as total volumes in compatible units.

Unfortunately, relationships such as eqn (12.2) do not necessarily allow for the increase in sediment production caused by urbanisation. Curtis and McCuen (1977) have therefore proposed a modelling approach in which the detachment and transport of soil by both rainfall and runoff is based upon relationships given by Meyer and Wischmeier (1969), and sediment removal within the pond is simulated using the theory proposed by Camp (1945) for the design of settling tanks for water treatment. For ponds which are to be used specifically for sediment control during urban development, Yrjanainon (1975) has suggested a simplified approach in which the settling velocity of the average size of sediment particle is equated to the velocity of flow through the pond. The latter is given by the quotient of the peak inflow rate and the surface area of the pond.

As an alternative to modelling the processes of erosion and sedimentation, several authors have developed regression equations for trap efficiency based upon measurements of the inflows and outflows to individual ponds and reservoirs. Results presented by McCuen (1980) showed that TE could be related to the extent of impervious area, the volume of inflow and the duration of rainfall excess. In a similar study by Rausch and Schreiber (1981), the highest explained variance for sediment TE was obtained using detention time, total storm runoff and mean discharge weighted sediment concentration as the independent variables.

Although an increasing number of investigations have been mounted in which the sediment balance of various ponds, lakes and reservoirs has been monitored, as noted by Whipple (1979), few data are available on the trap efficiencies of different pollutants. The results obtained by Oliver and Grigoropoulos (1981) and Rausch and Schreiber (1981) are therefore of particular interest. The former authors carried out a 7-month study of a 2·3 ha recreational lake in Missouri with an average detention time of 28 days. Over this period, the trap efficiencies for total suspended solids, total phosphorus, chemical oxygen demand and organic nitrogen were found to be 89, 65, 52 and 31% respectively. The study by Rausch and Schreiber (1981) of an 8·8 ha flood storage pond again in Missouri was directed primarily at the factors affecting the TE for nutrients. Over a 3-year period the pond trapped an average of 85% of the incoming sediment, 77% of the total sediment phosphorus and 37% of the inorganic nitrogen. The average trap efficiencies of the readily available nutrients, such as nitrate, were about 40%. However, those authors warn that this performance may not be maintained as sediment continues to accumulate in the pond.

## REFERENCES


Aronson, D. A. and Prill, R. C. (1977) Analysis of the recharge potential of storm-water basins on Long Island, New York. *J. Res. U.S. Geol. Survey*, **5,** 307–18.

Barton, B. A. (1977) Short-term effects of highway construction on the limnology of a small stream in southern Ontario. *Freshwater Biol.*, **7,** 99–108.

Bonham, A. J. (1974a) Storm drainage system design and new city planning. *Civ. Engng. Trans., Instn. Engrs. Austr.*, **CE16**(1), 67–70.

Bonham, A. J. (1974b) Urban stormwater drainage planning and environmental design. *J. Roy. Austr. Plan. Inst.*, **12**(3), 86–9.

Brune, G. M. (1953) Trap efficiency of reservoirs. *Trans. Am. Geophys. Un.*, **34,** 407–18.

Bunyan, J. (1975) The development of a flood storage area at Basildon. *J. Instn. Wat. Engrs. Sci.*, **29,** 175—82.

Burton, K. R. (1980) Stormwater detention basin sizing. *Proc. Am. Soc. Civ. Engrs., J. Hydraul. Div.*, **106**(HY3), 437–9.

Burton, T. M., Turner, R. R. and Harriss, R. C. (1976) The impact of highway construction on a north Florida watershed. *Wat. Resour. Bull.*, **12,** 529–38.

Butler, R. J. M. (1972) Water as an unwanted commodity: some aspects of flood alleviation. *J. Instn. Wat. Engrs.*, **26,** 311–21.

Camp, T. R. (1945) Sedimentation and the design of settling tanks. *Proc. Am. Soc. Civ. Engrs.*, **71,** 445–86.

Cherkauer, D. S. (1975) Urbanisation impact on water quality during a flood in small watersheds. *Wat. Resour. Bull.*, **11,** 987–98.

Copas, B. S. (1957) Storm water storage calculations. *J. Instn. Pub. Health Engrs.*, **56**(3), 137–62.

Curtis, D. C. and McCuen, R. H. (1977) Design efficiencies of stormwater detention basins. *Proc. Am. Soc. Civ. Engrs., J. Wat. Resour. Plan. and Man. Div.*, **103**(WR1), 125–40.

Davis, L. H. (1963) The hydraulic design of balancing tanks and river storage pounds. *Chart. Munic. Engr.*, **90,** 1–7.

Davis, L. H. and Woods, D. R. (1979) Design and construction of balancing lakes at Milton Keynes. *Chart. Munic. Engr.*, **106,** 9–17.

Debo, T. N. and Ruby, H. (1982) Detention basins—an urban experience. *Pub. Wks.*, **113**(1), 42–3; 93–4.

Donahue, J. R., McCuen, R. H. and Bondelid, T. R. (1981) Comparison of detention basin planning and design models. *Proc. Am. Soc. Civ. Engrs., J. Wat. Resour. Plan. and Man. Div.*, **107**(WR2), 385–400.

Field, R., Struzewski, E. J., Masters, H. E. and Tafuri, A. N. (1973) Water pollution and associated effects from street salting; Water pollution control in low-density areas. *Proc. Rural Environ. Engng. Conf.* (Univ. Press of New England, Hanover, NH) pp. 317–340.

GRAF, W. L. (1975) The impact of suburbanisation on fluvial geomorphology. *Wat. Resour. Res.*, **11,** 690–2.

HALL, M. J. (1981) A dimensionless unit hydrograph for urbanising catchment areas. *Proc. Instn. Civ. Engrs., Part 2,* **71,** 37–50.

HALL, M. J. and HOCKIN, D. L. (1980) Guide to the design of storage ponds for flood control in partly urbanised catchment areas. Construction Industry Research and Information Association, Tech. Note 100, 103 pp.

HALL, M. J., PRUS-CHACINSKI, T. M. and RIDDELL, K. J. (1978) Some aspects of the design of stormwater balancing ponds for catchment areas subject to urbanisation. In Helliwell, P. R. (ed.), *Urban storm drainage*, Proc. Int. Conf., Southampton (Pentech Press, London) pp. 421–33.

HAMMER, T. R. (1972) Stream channel enlargement due to urbanisation. *Wat. Resour. Res.*, **8,** 1530–40.

HAWKINS, R. H. (1976) Salt storage and run-off in urban watershed. *Proc. Am. Soc. Civ. Engrs., J. Environ. Engng. Div.*, **102**(EE4), 737–43.

HAWLEY, M. E., BONDELID, T. R. and MCCUEN, R. H. (1981) A planning method for evaluating downstream effects of detention basins. *Wat. Resour. Bull.*, **17,** 806–13.

HEINEMANN, H. G. (1981) A new sediment trap efficiency curve for small reservoirs. *Wat. Resour. Bull.*, **17,** 825–30.

HOLLIS, G. E. and LUCKETT, J. K. (1976) The response of natural river channels to urbanisation: two case studies from southeast England. *J. Hydrol.*, **30,** 351–63.

JACKSON, T. J. and RAGAN, R. M. (1974) Hydrology of porous pavement parking lots. *Proc. Am. Soc. Civ. Engrs., J. Hydraul. Div.*, **100**(HY12), 1739–52.

KARR, J. R. and SCHLOSSER, I. J. (1978) Water resources and the land–water interface. *Science*, **201,** 229–34.

KELLER, E. A. and HOFFMAN, E. K. (1976) A sensible alternative to stream channelisation. *Pub. Wks.*, **107**(10), 70–2.

KELLER, F. J. (1962) Effect of urban growth on sediment discharge, Northwest Branch Anacostia River basin, Maryland. US Geol. Survey, Prof. Paper 450-C, pp. C129–C131.

KELLY, H. G. (1977) Designing retention basins for small land developments. *Wat. and Sewage Wks.*, **124**(10), 78–80.

LEOPOLD, L. B., WOLMAN, M. G. and MILLER, J. P. (1964) *Fluvial processes in geomorphology* (W. H. Freeman, San Francisco) 522 pp.

MCCUEN, R. H. (1974) A regional approach to urban storm water detention. *Geophys. Res. Lett.*, **1,** 321–2.

MCCUEN, R. H. (1979) Downstream effects of stormwater management basins. *Proc. Am. Soc. Civ. Engrs., J. Hydraul. Div.*, **105**(HY11), 1343–56.

MCCUEN, R. H. (1980) Water quality trap efficiency of storm water management basins. *Wat. Resour. Bull.*, **16,** 15–21.

MAYS, L. W. and BEDIENT, P. B. (1982) Model for optimum size and location of detention. *Proc. Am. Soc. Civ. Engrs., J. Wat. Resour. Plan. and Man. Div.*, **108**(WR3), 270–85.

MEIN, R. G. (1980) Analysis of detention basis systems. *Wat. Resour. Bull.*, **16,** 824–9.

MEIN, R. G. and WOODHOUSE, M. P. (1977) Design of retarding basin

systems. In: *The Hydrology of Northern Australia*, Proc. Hydrol. Symp., Instn. Engrs. Austr., Brisbane (Instn. Engrs. Austr., Sydney) pp. 141–145.
MEYER, L. D. and WISCHMEIER, W. H. (1969) Mathematical simulation of the process of soil erosion by water. *Trans. Am. Soc. Agric. Engrs.*, **12,** 754–62.
NATURAL ENVIRONMENT RESEARCH COUNCIL (1975) *Flood Studies Report*, Vol. I: *Hydrological studies* (NERC, London) 550 pp.
OLIVER, L. J. and GRIGOROPOULOS, S. G. (1981) Control of storm-generated pollution using a small urban lake. *J. Wat. Poll. Contr. Fed.*, **53,** 594–603.
RAUSCH, D. L. and SCHREIBER, J. D. (1981) Sediment and nutrient trap efficiency of a small flood-detention reservoir. *J. Environ. Qual.*, **10,** 288–93.
ROTH, D. and WALL, G. (1976) Environmental effects of highway de-icing salts. *Ground Water*, **14,** 286–9.
SARTOR, J. D., BOYD, G. B. and AGARDY, F. J. (1974) Water pollution aspects of street surface contaminants. *J. Wat. Poll. Contr. Fed.*, **46,** 458–67.
SCOTT, W. S. (1976) The effect of road de-icing salts on sodium concentration in an urban watercourse. *Environ. Poll.*, **10,** 141–53.
SEABURN, G. E. (1970) Preliminary results of hydrologic studies at two recharge basins on Long Island, New York. US Geol. Survey, Prof. Paper 627-C, 17 pp.
SEABURN, G. E. and ARONSON, D. A. (1974) Influence of recharge basins on the hydrology of Nassau and Suffolk Counties, Long Island, New York. US Geol. Survey, Water-Supply Paper 2031, 66 pp.
TALHAMI, A. (1980) Temporary detention cuts storm flow peaks. *Civ. Engng. (ASCE)*, **50**(12), 72–5.
TAPP, J. S., WARD, A. D. and BARFIELD, B. J. (1982) Approximate sizing of reservoirs for detention time. *Proc. Am. Soc. Civ. Engrs., J. Hydraul. Div.*, **108**(HY1), 17–23.
VICE, R. B., GUY, H. P. and FERGUSON, G. E. (1969) Sediment movement in an area of suburban highway construction, Scott Run basin, Fairfax County, Virginia, 1961–4. US Geol. Survey, Water-Supply Paper 1591-E, 41 pp.
WALLING, D. E. and GREGORY, K. J. (1970) The measurement of the effects of building construction on drainage basin dynamics. *J. Hydrol.*, **11,** 128–44.
WATSON, M. D. (1981) Sizing of urban flood control ponds. *Civ. Engr. in S. Africa*, **23,** 183–9.
WEBER, W. G. and REED, L. A. (1976) Sediment runoff during highway construction. *Civ. Engng. (ASCE)*, **46**(3), 76–9.
WHIPPLE, W. (1979) Dual-purpose detention basins. *Proc. Am. Soc. Civ. Engrs., J. Wat. Resour. Plan. and Man. Div.*, **105**(WR2), 403–12.
WOLMAN, M. G. and SCHICK, A. P. (1967) Effects of construction on fluvial sediment, urban and suburban areas of Maryland. *Wat. Resour. Res.*, **3,** 451–64.
WULKOWICZ, G. M. and SALEEM, Z. A. (1974) Chloride balance of an urban basin in the Chicago area. *Wat. Resour. Res.*, **10,** 974–82.
WYCOFF, R. L. and SINGH, U. P. (1976) Preliminary design of small flood detention reservoirs. *Wat. Resour. Bull.*, **12,** 337–49.
YRJANAINON, G. (1975) Sediment basin design. *Wat. and Sewage Wks.*, **122**(7), 82–4.

# Author Index

*Numbers in italic type indicate those pages on which references are given in full*

# Subject Index

LADY MARGARET HALL LIBRARY
OXFORD